Annals of Mathematics Studies

Number 134

Temperley-Lieb Recoupling Theory and Invariants of 3-Manifolds

by

Louis H. Kauffman and Sóstenes L. Lins

PRINCETON UNIVERSITY PRESS

———

PRINCETON, NEW JERSEY

1994

The Annals of Mathematics Studies are edited by
Luis A. Caffarelli, John N. Mather, and Elias M. Stein

Princeton University Press books are printed on acid-free paper and meet the
guidelines for permanence and durability of the Committee on Production
Guidelines for Book Longevity of the Council on Library Resources

Printed in the United States of America

10 9 8 7 6 5 4 3 2 1

Library of Congress Cataloging-in-Publication Data

Kauffman, Louis, H., 1945–
  Temperley-Lieb recoupling theory and invariants of 3-manifolds /
by Louis H. Kauffman and Sóstenes L. Lins.
    p.   cm. — (Annals of mathematics studies ; no. 134)
  Includes bibliographical references and index.
  ISBN 0-691-03641-1   ISBN 0-691-03640-3 (pbk.)
    1. Knot theory. 2. Three-manifolds (Topology).
3. Invariants (Mathematics)   I. Sóstenes L. Lins. II.Title.
III. Series.
QA612.2.K39   1994
514'.224—dc20   94-15512

The publisher would like to acknowledge the authors of this volume for providing
the camera-ready copy from which this book was printed

Contents

1 Introduction   1

2 Bracket Polynomial, Temperley-Lieb Algebra   5
   2.1 Bracket Polynomial   5
   2.2 Temperley-Lieb Algebra   8

3 Jones-Wenzl Projectors   13
   3.1 A Standard Projector in T_n   13
   3.2 The Projectors as Sums of Tangles   15
   3.3 Diagrams and Structural Recursion   18

4 The 3-Vertex   22
   4.1 A Special Sum of Tangles   22
   4.2 A Fundamental Twist   24
   4.3 Invariants of Trivalent Embedded Graphs: $P_a(G)$ 30
   4.4 The Case of $P_2(G)$ 31

5 Properties of Projectors and 3-Vertices   36
   5.1 Vanishing Conditions   36
   5.2 Interaction with Curls and Loops   42

6 θ-Evaluations   45
   6.1 Recursive Relations   45
   6.2 Shaping the Recursion   51
   6.3 A Formula for the θ-nets 55

7 Recoupling Theory Via Temperley-Lieb Algebra   **60**

7.1 Recoupling Theorem   60

7.2 The case of General q   66

7.3 Orthogonality and Pentagon Identities   69

8 Chromatic Evaluations and the Tetrahedron   **76**

8.1 Exact Formulas in a Special Case   76

8.2 Tensorial Formalism   77

8.3 A Heuristic Correspondence on the θ-Net   81

8.4 Chromatic Evaluation: General Case   83

8.5 The Tetrahedron   88

9 A Summary of Recoupling Theory   **93**

9.1 Bracket Polynomial   93

9.2 Temperley-Lieb Algebra T_n   94

9.3 Chebyschev Polynomials   94

9.4 Quantum Integers   95

9.5 q-Symmetrizer   95

9.6 Jones-Wenzl Projectors   96

9.7 Curl and Projector   96

9.8 Loop and Projector   96

9.9 3-Vertex   97

9.10 θ-Net   97

9.11 Tetrahedral Net   98

9.12 q-$6j$ Symbols   99

9.13 Orthogonality Identity   99

9.14 Biedenharn-Elliot (Pentagon) Identity   99

9.15 Two Special Cases   99

9.16 Axiomatics   100

10 A 3-Manifold Invariant by State Summation   **102**

10.1 Matveev-Piergallini Moves   102

10.2 A Partition Function   104

10.3 Invariance under Lune Move   108

10.4 Invariance under the Y-move   109

10.5 Behavior under Bubble Move   112

11 The Shadow World — 114

11.1 Preliminaries 114
11.2 Shadow Translations 116
11.3 Proving Shadow World Transitions 120
11.4 Examples 125

12 The Witten-Reshetikhin-Turaev Invariant — 129

12.1 Framed Links 129
12.2 Examples 131
12.3 Handle Sliding and Kirby Calculus 133
12.4 Consequences of Handle Slides 135
12.5 Invariants 140
12.6 Lickorish's Proof 144
12.7 Normalization 146
12.8 Gauss Sums 150
12.9 Examples 152
12.10 Shadow Interpretation 153
12.11 Appendix: Invariants of 4-Manifolds 156

13 Blinks $\mapsto$ 3-Gems: Recognizing 3-Manifolds — 160

13.1 Motivating 3-Gems 160
13.2 Graph-Encoded 3-Manifolds 163
13.3 Dipole Moves: Ferri-Gagliardi Theorem 167
13.4 The Sufficiency of the Matveev-Piergallini Moves 168
13.5 From Blinks to 3-Gems 171
13.6 Rigid 3-Gems 175
13.7 The Code of a Bipartite $(n+1)$-Graph 178
13.8 TS-Classes: a Basis for 3-Manifold Classification 180

14 Tables of Quantum Invariants — 185

14.1 Overview of the Tables 185
14.2 Knot 3_1 191
14.3 Knot 4_1 205
14.4 Knot 5_1 212
14.5 Knot 5_2 227
14.6 Knot 6_1 235
14.7 Knot 6_2 239
14.8 Knot 6_3 243

14.9 Knot 7_1 245

14.10 Knot 7_2 249

14.11 Knot 7_3 253

14.12 Knot 7_4 257

14.13 Knot 7_5 261

14.14 Knot 7_6 265

14.15 Knot 7_7 269

14.16 Links with 2 Components 272

14.17 Links with 3 Components 286

Bibliography   **290**

Index   **295**

Temperly-Lieb Recoupling Theory
and Invariants of 3-Manifolds

Chapter 1

Introduction

This monograph develops, in a self-contained manner, a recoupling theory for colored knots and links with trivalent graphical vertices. The theory is based on underlying properties of the bracket polynomial and the tangle-theoretic Temperley-Lieb algebra ([Kau87b],[Kau90b]). This recoupling theory is the direct analog, in this context, of the corresponding theory for q-deformed angular momentum recoupling using the quantum group $SL(2)_q$ (See [KR88].).

By working directly with the knot theory associated with the bracket polynomial we obtain a direct and combinatorial approach to this subject. In particular, we obtain a direct analysis of the recoupling theory at roots of unity that is useful for applications to topological invariants of 3-manifolds. In particular, we obtain an essentially axiomatic reconstruction of the invariant of Turaev and Viro (Chapter 10) and a corresponding treatment of the Witten-Reshetikhin-Turaev invariant in Chapter 12.

An important motivation for this work was the realization that the Temperley-Lieb context can be viewed as a q-deformed version of the spin-network theory of Roger Penrose [Pen69]. This point of view is not detailed here, but can be found in other papers by the first author ([Kau90c][Kau91] [Kau90a] [Kau92]).

In Chapter 8 we use a combination of the chromatic technique of Penrose and Mousourris [Mou79] and q analogs to obtain the basic formulas for recoupling coefficients needed in this theory. Further connections with spin networks and with mathematical physics will be given in a separate paper.

1

The book is organized as follows. Chapter 2 discusses the bracket polynomial model of the original Jones polynomial and its relation with the Temperley-Lieb algebra. Chapter 3 discusses a definition of Jones-Wenzl projectors in the Temperley-Lieb algebra as q-deformed symmetrizers. This construction is conceptually interesting, and it is computationally useful for the chromatic method explained in Chapter 8. Chapter 4 goes on to define the 3-vertex in terms of these projectors and ends with a discussion of examples and of n-fold cabled bracket invariants. Chapter 5 and Chapter 6 take up the computation of θ-networks at roots of unity and for generic values of q. ($\sqrt{q} = A$ where A is the variable in the bracket polynomial.) In the course of this discussion a number of basic formulas for network evaluation are derived. In Chapter 7 we do the recoupling theory both for generic q and at roots of unity. This can be summarized via formulas of the form

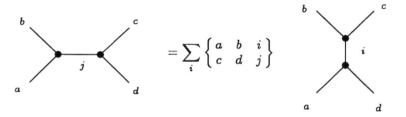

where the indices are subject to certain admissibility conditions, equality refers to global network evaluation and the symbol $\left\{ \begin{matrix} a & b & i \\ c & d & j \end{matrix} \right\}$ is a (generalized) $6j$ symbol receiving its definition from the above formula (see Chapter 7). Propositions 9 and 10 prove the orthogonality and the Biedenharn-Elliot identities for these $6j$ symbols. Proposition 11 gives our network explicit formula for $6j$:

Analysis of the θ-nets in Chapter 6 guarantees that, for q-admissible labellings when q is a root of unity, these formulas are well defined and correspond to labels in the set $\{0, 1, 2, \ldots, r - 2\}$ for $q = e^{i\pi/r}$. Chapter 8 contains the computations of the theta and tetrahedral coefficients via counting formulas in the classical case and recursions in the q-deformed case. Chapter 10 details our construction of a 3-manifold invariant by state summation, using this recoupling theory. This invariant is identical with the invariant of Turaev and Viro [TV92] (See [Piu92]), and our construction of it shows how the invariance follows from axiomatic features of the network recoupling. The state sum (Definition 7) that defines the invariant takes on a particularly transparent form when given in terms of a labelling of the decomposition of special spine for a 3-manifold M^3.

Chapter 9 contains a succinct summary of the facts of the recoupling theory. The reader may find it useful to go directly to it and beyond after reading Chapters 2,3 and 4. After discussing the Turaev-Viro invariant in Chapter 10, we turn to the Shadow World of Kirillov and Reshetikhin in Chapter 11. Here we show how to translate colored link invariants into partition functions of very similar type to the Turaev-Viro invariant. This is accomplished by a careful reformulation of the Kirillov-Reshetikhin work in terms of our recoupling theory. In Chapter 12 we construct the Witten-Reshetikhin-Turaev invariant, give a normalization for it at $A = e^{i\pi/2r}$ and then use the results of Chapter 11 to reformulate this invariant as a partition function on a 2-complex. The Chapter ends with a sketch of the proof of the basic Theorem of Turaev and Walker relating these invariants.

An appendix to Chapter 12 sketches the work of Justin Roberts on the Turaev-Viro invariant and its application to the Crane-Yetter invariant of 4-manifolds.

A *blink* is a finite plane graph with a bipartition on the edges. Blinks are in $1 - 1$ correspondence with projections of links and so, via the blackboard framing, they are very concise data structure which yield connected orientable 3-manifolds. This presentation of 3-manifolds is entirely adequate for the computation of the Witten-Reshetikhin-Turaev invariants, as we show in Chapters $12-14$. One disadvantage of the presentation by blinks is that the recognition of the 3-manifolds from them is rather difficult. This is due to the lack of a useful simplification theory. In Chapter 13 we provide a simple way to go from a blink to a particular kind of ball complex inducing the same 3-manifold. These ball complexes are special 4-valent edge colored graphs named 3-*gems*. These objects do have a useful simplification theory which we survey in Chapter 13. In particular, 3-manifolds induced by 3-gems with less than 30 vertices have been topologically classified. With the simplification theory of 3-gems we recognize many of the 3-manifolds for which we give the Witten-Reshetikhin-Turaev invariants in Chapter 14. Various pairs of 3-manifolds in these tables are suspected to be homeomorphic from these invariants. They are proven to be so because they have the *same TS-code:* a canonical form computed in 3-gems (see Chapter 13). We stress the point that the implementation of the algorithm to get, as in Chapter 14, the quantum invariants for 3-manifolds from blinks is based solely on well known basic results on framed links and, otherwise, on results developed from scratch, in this monograph. Fully detailed recipes on how to reproduce these computations are discussed and crystallized.

It gives us pleasure to thank Vladimir Turaev, Ray Lickorish, Oleg Viro and Said Sidki for helpful conversations in the course of this research. The implementation of the TS_ρ-algorithm, from which we recognize various manifolds, in the tables of Chapter 14 is due to Cassiano Durand. We also thank Oscar Pereira S. Neto for the latex typing and elaboration of most of the hundreds of figures, some of them rather intricate.

The first author wishes to thank the National Science Foundation for support under NSF Grant Number DMS-9205277 and the Program for Mathematics and Molecular Biology at the University of California at Berkeley for partial support during the preparation of this research. He also thanks the Research Institute for Mathematical Sciences, Kyoto University, Kyoto, Japan; the Isaac Newton Institute, Cambridge, England and the Universidade Federal de Pernambuco in Recife, Brazil for their kind hospitality during the course of this work.

The second author wants to acknowledge the partial support of FINEP and CNPq/Brazil (process number 30.1103/80). He also thanks the University of Illinois at Chigago and the Geometry Center at Minneapolis, Institutions where part of this work were produced.

Chapter 2

Bracket Polynomial, Temperley-Lieb Algebra

2.1 Bracket Polynomial

We start by recalling the definition and elementary properties of the (topological) bracket polynomial: to each crossing in an unoriented link diagram K there are associated two smoothings, labelled A and A^{-1} as shown in Figure 1.

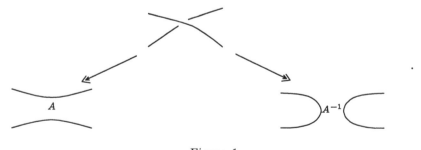

Figure 1

Labellings for these two types of smoothing are obtained by labelling the four regions at the crossing by A and A^{-1} so that the two regions swept out by turning the over-crossing line counter-clockwise are labelled A. This configuration is shown below.

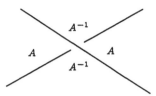

The smoothing that joins the two A-labelled regions is the A-smoothing. The smoothing that joins the two A^{-1}-labelled regions is the A^{-1}-smoothing.

A *state S of the diagram K* is a choice of a smoothing for each crossing in K. Thus S appears as a disjoint set of Jordan curves in the plane that is decorated with labels at the site of each smoothing. For example, the diagram below is a state of the trefoil diagram K

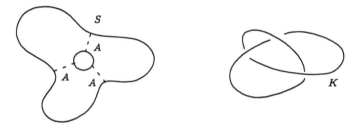

Given a state S of a diagram K, let $||S||$ denote the number of disjoint Jordan curves in S, and let $< K|S >$ denote the product of the state labels (called vertex weights in the statistical mechanics context [Kau89]) of S. (In the example above, $||S|| = 2$ and $< K|S >= A^3$.)

Define the *bracket polynomial* $< K >\in \mathbb{Z}[A, A^{-1}]$ by the state summation formula,

$$< K >= \sum_{S} < K|S > d^{||S||}$$

where S runs all states of the diagram K, and $d = -A^2 - A^{-2}$. The following results are proved in [Kau85], [Kau87b], [Kau87a], (see also [Kau83].).

Theorem 1 *The bracket polynomial is an invariant of regular isotopy of link diagrams. If K is an oriented link diagram with $w(K)$ the writhe of K, then*

$$f_K(A) = (-A^3)^{-w(K)} < K > / < 0 >$$

is an invariant of ambient isotopy of link diagrams. The bracket polynomial has the following properties.

(i) $\left\langle \asymp \right\rangle = A \left\langle \smile\frown \right\rangle + A^{-1} \left\langle \supset\subset \right\rangle$

where the small diagrams stand for parts of larger ones that differ only as indicated by them.

(ii) $< 0 \sqcup K >= d < K >$
where $0\sqcup$ denotes disjoint union of the diagram K with a Jordan curve in the plane, and $d = -A^2 - A^{-2}$.

(iii) If $V_K(t)$ is the original Jones polynomial [Jon86], then

$$V_K(t) = f_K(t^{-1/4}).$$

These results are all elementary consequences of the definition of the bracket polynomial. Recall, however the definitions of regular isotopy, ambient isotopy and writhe. Two link diagrams are said to be ambient isotopic if one can be obtained from the other by a sequence of Reidemeister moves of the type I, II, III (plus underlying graphical changes induced by homeomorphisms of the plane). Two link diagrams are regularly isotopic if one can be obtained from the other by a sequence of Reidemeister moves of type II and III. The Reidemeister moves are shown in Figure 2.

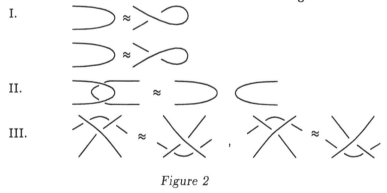

Figure 2

The *writhe*, $w(K)$, of an oriented link diagram K is the sum of the signs of the crossings in the diagrams, where these signs are given by the convention shown below.

$$\varepsilon = +1 \qquad \varepsilon = -1 \qquad w(K) = \sum_p \varepsilon(p).$$

2.2   Temperley-Lieb Algebra

The Jones polynomial [Jon86] was originally defined via a representation of the Artin braid group [Art25] into the Temperley-Lieb algebra ([Jon87], [Jon83]). The bracket model for Jones polynomial allows a tangle-theoretic interpretation of the Temperley-Lieb algebra (see [Kau85], [Kau87b]) that is fundamental to the rest of the work in this paper.

In order to elucidate this version of Temperley-Lieb algebra, we define the *elementary tangles* $U_1, U_2, \ldots, U_{n-1} \in T_n$ where T_n denotes the n-strand Temperley-Lieb algebra. That is, each U_i is a tangle with n input strands and n output strands. In U_i, the kth input is connected to the kth output for $k \neq i, i+1$, while the ith input is connected to the $(i+1)th$ input and the ith output is connected to the $(i+1)th$ output. See Figure 3 for an illustration of T_4.

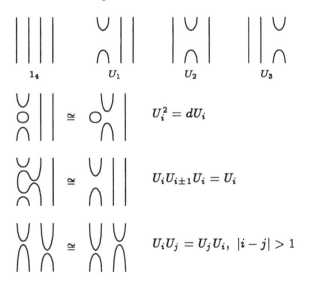

$$1_4 \qquad U_1 \qquad U_2 \qquad U_3$$

$$U_i^2 = dU_i$$

$$U_i U_{i\pm 1} U_i = U_i$$

$$U_i U_j = U_j U_i, \ |i - j| > 1$$

Figure 3

Two tangles of the same number of strands are multiplied by attaching the output strands of the first tangle to the input strands of the second tangle. Two tangles are *equivalent* if they are regularly isotopic relative to their end points. This means that the regular isotopy of a given tangle is restricted to a box; the input and output strands emanate from this box; the regular isotopy leaves the intersections of the input and output strands with the boundary of the box fixed.

It is not hard to see that the equivalence class of a product of the $U_i's$ is determined entirely by the pattern of connections of inputs and outputs and that these products include all possible such connections that can be drawn in a planar box without self intersections. For example, in T_3,

$$\cong \quad = U_2 U_1.$$

The $U_i's$ enjoy the following basic relations (Compare Figure 3):

(i) $U_i^2 = dU_i$ where d denotes a (commuting) value assigned to a closed loop.

(ii) $U_i U_{i\pm1} U_i = U_i$

(iii) $U_i U_j = U_j U_i$ for $|i - j| > 1$.

Call a tangle *planar-non-intersecting* if it can be represented in the plane with no intersections and no crossings. In [Kau90b] it is shown that every planar non-intersecting n-tangle (n inputs and n outputs) is equivalent to a product of $1_n, U_1, U_2, \ldots, U_{n-1}$ (1_n is the tangle that connects the i^{+h} input with the i^{+h} output) and that two such products represent equivalent tangles if and only if one product can be obtained from the other by the relations (i), (ii), (iii) above.

The *Temperley-Lieb algebra* T_n is the free additive algebra over $\tilde{\mathbb{Z}}[A, A^{-1}]$ with multiplicative generators $1_n, U_1, U_2, \ldots, U_{n-1}$ and relations (i), (ii) and (iii) as given above. We take $d = -A^2 - A^{-2}$, and it is given that A and A^{-1} commute with all elements of T_n. [Here $\tilde{\mathbb{Z}}[A, A^{-1}]$ denotes the set of rational functions P/Q with $P, Q \in \mathbb{Z}[A, A^{-1}]$].

Remark 1 *For our purposes it is convenient to parametrize the loop* d *as* $d = -A^2 - A^{-2}$. *Other treatments of Temperley-Lieb algebra use different loop parametrizations.*

It is useful to generalize the Temperley-Lieb algebra to an algebra T_n that is generated multiplicatively by the regular isotopy equivalence classes of all n-strand tangles. Thus a generator of T_n is any link diagram with $2n$ free ends such that n of them are designated as inputs and the remaining n of them are designated as outputs. The diagram is regarded as enclosed in a box from which emanate the inputs and the outputs. The *tangle algebra* T_n is the free additive algebra over $\tilde{\mathbb{Z}}[A, A^{-1}]$ generated multiplicatively by the n-strand tangles (n-tangles).

If x is an n-tangle, let $\bar{x}$ denote the *standard closure of* x obtained by attaching the kth input to the kth output as shown below.

Now define the *trace*
$$tr : T_r \longrightarrow \mathbb{Z}[A, A^{-1}]$$
by the formulas

(i) If x is an n-tangle then $tr(x) = < \bar{x} >$ where $< >$ denotes the bracket polynomial.

(ii) $tr(x + y) = tr(x) + tr(y)$.

Note that $tr(ab) = tr(ba)$ is an immediate consequence of the properties of the bracket polynomial and of the form of closure for the tangles.

Since the Temperley-Lieb algebra is a sub-algebra of the tangle algebra, this trace function restricts to a trace on T_n. In this case, if $x \in T_n$ is a product of the $U_i's$, then $\bar{x}$ is a disjoint union of Jordan curves in the plane, and $tr(x) = < \bar{x} > = d^{||\bar{x}||}$ where $||\bar{x}||$ denotes the number of Jordan curves. Each product of $U_i's$ corresponds to a single bracket state.

Remark 2 *This trace on the Temperley-Lieb algebra corresponds directly to the Jones trace [Jon83]. In [Jon83] this trace is defined in terms of normal forms for words in the algebra.*

Elements of the Artin braid group B_n are a special case of n-tangles. A *braid* $b \in B_n$ is a n-tangle that is regularly isotopic to a product of *elementary braids* $1_n, \sigma_1, \ldots, \sigma_{n-1}, \sigma_1^{-1}, \ldots, \sigma_{n-1}^{-1}$. The elementary braid $\sigma_i^{\pm}$ takes input i to output $i+1$ and input $i+1$ to output i. The braids σ_i and σ_i^{-1} have opposite crossing type, with σ_i such that the smoothing that joins input to input and output to output is A-type. (See Figure 4.) Call this the *horizontal smoothing* of σ_i and denote it by $H(\sigma_i)$. Similarly, let $V(\sigma_i)$ denote the input-output (vertical) smoothing of σ_i. Thus we have

$$H(\sigma_i^{\pm}) = U_i \, , \qquad V(\sigma_i^{\pm}) = 1_n.$$

Since a state (bracket state) of the closure $\bar{b}$ of a braid b is obtained by choosing a smoothing for each $\sigma_i^{\pm}$ in b, it follows that *each state of $\bar{b}$ corresponds to the strand closure of an element in the Temperley-Lieb algebra.*

For example, the states of $\bar{\sigma}_1$ in B_2 are $\overline{H(\sigma_1)} = \overline{U_1}$ and $\overline{V(\sigma_1)} = \overline{1_2}$. We have

$$< \bar{\sigma}_1 >= Ad^{||\overline{U_i}||} + A^{-1}d^{||\overline{1_2}||} = Ad + A^{-1}d^2 \quad (d = -A^2 - A^{-2}).$$

This gives an algebraic algorithm for computing $< \bar{b} >$ for any braid b via a sum of trace evaluations of elements of the Temperley-Lieb algebra. In fact the method applies with slight generalization to any tangle, but in the case of the braid group there is an underlying representation $\rho : B_n \longrightarrow T_n$ of the braid group to the Temperley-Lieb algebra. This representation is determined by the formulas

$$\rho(\sigma_i) = AU_i + A^{-1}1_n$$

$$\rho(\sigma_i^{-1}) = A^{-1}U_i + A1_n.$$

It is easy to check [Kau87b],[Kau90b] that with loop value $d = -A^2 - A^{-2}$ that ρ is a representation is equivalent to the representation constructed by Jones [Jon83], and it follows from our definitions that

$$tr\rho(b) =< \bar{b} >,$$

giving the bracket as a trace on the representation of the braid group into the Temperley-Lieb algebra.

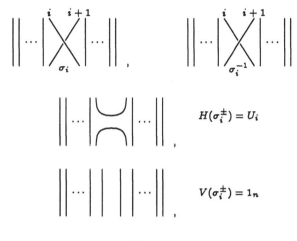

$$H(\sigma_i^{\pm}) = U_i$$

$$V(\sigma_i^{\pm}) = 1_n$$

Figure 4

Chapter 3

Jones-Wenzl Projectors

The next construction is motivated by a number of considerations, ranging for the need for graph invariants to the representation theory of the Temperley-Lieb algebra and an analogue of the theory of angular momentum.

3.1 A Standard Projector in T_n

Definition 1 *Let $f_i \in T_n$ be defined inductively for $i = 0, 1, 2, \ldots, n-1$ by the following formulas:*

$$
\begin{aligned}
f_0 &= 1_n \\
f_{k+1} &= f_k - \mu_{k+1} f_k U_{k+1} f_k
\end{aligned}
$$

where $\mu_1 = d^{-1}, \mu_{k+1} = (d - \mu_k)^{-1}$. Here d is the loop value in T_n, $d = -A^2 - A^{-2}$, and $U_i^2 = dU_i$ for each i.

Lemma 1 *The elements $f_i \in T_n$ enjoy the following properties.*

   (i) $f_i^2 = f_i$ *for $i = 0, 1, \ldots, n-1$.*

  (ii) $f_i U_j = U_j f_i = 0$ *for $j \geq i$.*

 (iii) $tr(f_{n-1}) = \Delta_n = \Delta_n(-A^2)$ *and $\mu_{k+1} = \Delta_k/\Delta_{k+1}$ with $\Delta_0 = 1$.*

where $\Delta_n(x) = \left(\dfrac{x^{n+1} - x^{-n-1}}{x - x^{-1}} \right)$ is the $n\underline{th}$ Chebyschev polynomial.

Proof: See [Jon83], [Lic91], or [Kau91].   ∎

Remark 3 *By adding one more input and one more output, and a strand between them, we have inclusions $T_1 \subset T_2 \subset T_3 \subset \cdots \subset T_n \subset T_{n+1} \subset \cdots$. In this sense, each $f_{i-1} \in T_i$ and we can take this as the usual placement of f_{i-1}. In this case, we can state that $tr(f_{i-1}) = \Delta_i$ (meaning the trace with respect to i-tangles).*

Proposition 1 *There is a unique non-zero element $f \in T_n$ such that*
(i) $f^2 = f$
(ii) $fU_i = U_i f = 0$, $i = 1, 2, \ldots, n-1$.

Proof: Lemma 1 asserts the existence of the element. Suppose that g also satisfies (i) and (ii). Since $g \in T_n$, g is a linear combination of products of the elements $U_1, U_2, \ldots, U_{n-1}$ and 1_n. Since $g^2 = g$, $g = 1_n + U$ where $g\,U = 0$, U being a sum of such products, each including at least one U_i for some i. Similarly $f = 1 + U'$. Hence

$$\begin{aligned} f &= f + fU = f(1 + U) = fg \\ &= (1 + U')g = g + U'g = g \end{aligned}$$

∎

Example: In T_2,

$$\begin{aligned} f_1 &= f_0 - \mu_1 f_0 U_1 f_0 \\ &= 1_2 - d^{-1} U_1. \end{aligned}$$

$$f_1 = \left|\;\right| - \frac{1}{d}\,\cup\!\!\cap$$

$$Tr(f_1) = \langle\!\langle \bigcirc\!\!\!\bigcirc \rangle\!\rangle - \tfrac{1}{d}\langle\!\langle \mathcal{S} \rangle\!\rangle$$

$$\begin{aligned} &= d^2 - (\tfrac{1}{d})d \\ &= d^2 - 1 \\ &= (-A^2 - A^{-2})^2 - 1 \\ &= A^4 + 1 + A^{-4}. \end{aligned}$$

$$\begin{aligned} \Delta_2(x) &= (x^3 - x^{-3})/(x - x^{-1}) \\ &= x^2 + 1 + x^{-2} \\ \therefore\quad tr(f_1) &= \Delta_2(-A^2). \end{aligned}$$

Proposition 1 is useful because it allows us to verify other realizations of these projectors. In particular, we now can give a global formula for f_{n-1}.

3.2 The Projectors as Sums of Tangles

For a given positive integer n, let $\boxed{}^{\,n}$ denote the sum (over rational functions in A and A^{-1} as coefficients) of n-tangles described below. Note that any n-tangle can be expended via the bracket identity

$$\asymp \;=\; A\,\smile\!\!\frown \;+\; A^{-1}\,)\,($$

to (a sum of) elements in the Temperley-Lieb algebra. Thus, we can regard $\boxed{}^{\,n}$ as in T_n. The definition is:

Definition 2

$$\boxed{}^{\,n} \;=\; \frac{1}{\{n\}!} \sum_{\sigma \in S_n} (A^{-3})^{t(\sigma)}\, \boxed{\hat{\sigma}}$$

where $\{n\}! = \sum_{\sigma \in S_n}(A^{-4})^{t(\sigma)} = \Pi_{k=1}^n\left(\frac{1-A^{-4k}}{1-A^{-4}}\right)$. Here S_n denotes the symmetric group on n letters, so that $\sigma \in S_n$ may be thought of as a permutation of $1, 2, \ldots, n$, and $\hat{\sigma}$ denotes the n-tangle obtained from any minimal representation of σ as a product of transpositions, so that each transposition is replaced by a braid in the form σ_i (note the positivity) for $i = 1, 2, \ldots, n-1$. For example, the permutation $a = \begin{pmatrix} 1 & 2 & 3 \\ 3 & 1 & 2 \end{pmatrix}$ corresponds to the braid $\hat{\sigma} = $ ⧓ .

The integer $t(\sigma)$ is the number of transpositions in the minimal representation of the permutation σ. (In the example above, $t(\sigma) = 2$.) The term $\{n\}!$ is a version of q-deformed factorial with $\sqrt{q} = A$. When $A = 1$ or -1, $\{n\}!$ is the usual factorial function.

Proposition 2 *If $g_n \in T_n$ denotes the image of* ⌐∏⌐ *in the Temperley-*

Lieb algebra T_n, then
  (i) $g_n^2 = g_n$
  (ii) $g_n U_i = U_i g_n = 0$ *for* $i = 1, 2, \ldots, n-1$.
Therefore, by Proposition 1 we conclude that $f_{n-1} = g_n$ in T_n. That is,

$$f_{n-1} = \;\; \boxed{}^{n}$$

Remark 4 *Before proving the Proposition 2, an example is in order. Here is the explicit expansion of g_2 in T_2.*

$$g_2 \;\; = \;\; \boxed{} \;\; = \;\; \frac{1}{\{2\}!}\left[\;\Big|\Big|\; + A^{-3}\;\asymp\;\right]$$

$$= \;\; \frac{1}{1+A^{-4}}\left[\;\Big|\Big|\; + A^{-3}\left[A\;\overset{\cup}{\cap}\; + A^{-1}\right)\Big(\Big|\Big]\right]$$

$$= \;\; \frac{1}{1+A^{-4}}\left[\left(1+A^{-4}\right)\Big|\Big|\; + A^{-2}\;\overset{\cup}{\cap}\right)\right]$$

$$= \;\; \Big|\Big|\; + \frac{1}{A^2 + A^{-2}}\;\overset{\cup}{\cap}$$

$$= \;\; f_1 \quad (\text{with } d = -A^2 - A^{-2}).$$

Note also that $gU_1 = 0$ via the following calculation

$$g_2 U_1 \;\; = \;\; \boxed{} \;\; = \;\; \frac{1}{\{2\}!}\left[\;\overset{\cup}{\cap}\; + A^{-3}\;\overset{\diagup}{\frown}\;\right]$$

$$= \;\; \frac{1}{\{2\}!}\left[\;\overset{\cup}{\cap}\; + A^{-3}(-A^3)\;\overset{\cup}{\cap}\;\right]$$

$$g_2 U_1 \;\; = \;\; 0.$$

Here we used $\bigvee\hspace{-1em}\bigcirc = (-A^3)\;\bigcup$, a fact about the bracket expansion.

As background for the proof of Proposition 2 it is useful to have a canonical inductive construction for the braids $\hat{\sigma}$ with $\sigma \in S_n$. To this end, note that by the convention of lifting to positive braids, the strand from input 1 to output i will *overcross* all the other strands that it meets. Hence the set $\{\hat{\sigma} | \sigma \in S_n\}$ is constructed from the set $\{\hat{\tau} | \tau \in S_{n-1}\}$ by inserting the braids $\hat{\tau}$ as inputs $2, 3, \ldots, n$ and outputs $1, 2, \ldots, i-1, i+1, \ldots, n$ with an overcrossing strand from input 1 to output i for each $i = 1, 2, \ldots, n$. For example, $\{\hat{\sigma} | \sigma \in S_3\}$ is constructed from $\{\hat{\tau} | \tau \in S_2\}$ by filling in the templates

as shown below.

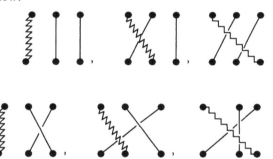

Call this the *canonical inductive construction* of the braid set $\{\hat{\sigma} | \sigma \in S_n\}$.

Proof: [of Proposition 2] It is easy to see from the canonical inductive construction of the braid set that $\{\hat{\sigma} | \sigma \in S_n\}$ can be written, for any given $i \in \{1, 2, \ldots, n-1\}$, as a disjoint union of a set of braids W and the set $W' = \{w\sigma_i | w \in W\}$. (Note that this entails the assumption that no braid word in W ends in σ_i or σ_i^{-1} due to minimality.) As a result, we have

$$\{n\}! g_n = \sum_{w \in W} (\mathcal{A}^{-3})^{t(w)} w + (\mathcal{A}^{-3})^{t(w)+1} w\sigma_i.$$

Since $w\sigma_i U_i = (-A^3) w U_i$ in T_n, it follows that $g_n U_i = 0$ for $i = 1, 2, \ldots, n-1$. This completes the proof of property (ii).

Next it is claimed that the coefficient of 1_n (the identify braid) in $\hat{g}_n = \sum_{\sigma \in S_n} (A^{-3})^{t(\sigma)} \hat{\sigma}$ is precisely $\{n\}!$, so that the coefficient of 1_n in

g_n is just equal to 1. To see this claim, note that for each $\hat{\sigma}$, its individual expansion in T_n contains one copy of 1_n. Hence each $\hat{\sigma}$ contributes $(A^{-3})^{t(\sigma)}(A^{-1})^{t(\sigma)}$ to the coefficient of 1_n. The factor $(A^{-1})^{t(\sigma)}$ is the product of vertex weights in the bracket expansion that yields the state 1_n. Hence the coefficient of 1_n in $\tilde{g}_n$ is the sum $\sum_{\sigma \in S_n}(A^{-4})^{t(\sigma)} = \{n\}!$, as claimed. Thus we know that g_n can be written in the form $g_n = 1_n + \mathcal{U}$ where $\mathcal{U}$ is a sum of products of $U_i's$. Therefore, $g_n^2 = g_n(1_n + \mathcal{U}) = g_n + g_u\mathcal{U} = g_n + 0 = g_n$ since $g_nU_i = 0$ for $i = 1, \ldots, n-1$. Thus $g_n^2 = g_n$, and the proof is complete.   ∎

The above result is useful since it shows that $\{n\}!f_{n-1}$ is an element of T_n all of whose coefficients are in $\mathbb{Z}[A, A^{-1}]$, hence all possibilities of poles in the coefficients of f_{n-1} are referred to questions about the zeroes of $\{n\}!$ (for specific complex values of A). It is also useful to use the tangle

notation $f_{n-1} = \boxed{}^{\,|n}$. Thus the recursion relation of definition 1 becomes

the diagrammatic equation

$$
\boxed{}^{\,|n+1} \quad = \quad \boxed{}^{\,|n}\Big| \quad - \quad \mu_n \; \boxed{}^{\,|n-1}
$$

with

$$
\begin{aligned}
\mu_1 &= 1/d \quad (d = -A^2 - A^{-2}) \\
\mu_{k+1} &= (d - \mu_k)^{-1}.
\end{aligned}
$$

3.3   Diagrams and Structural Recursion

The above diagrammatic recursion will be of fundamental use in proving our results on recoupling theory at roots of unity. That it is true follows from the uniqueness given in Proposition 1. A direct proof from the tangle definition of g_n is also possible.

The diagrammatic notation makes it easy to see how this recursive

relation is structurally determined. Assume that $g_n = \boxed{}^{\,|n} \in T_n$ is given

and that

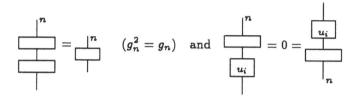

$$(g_n^2 = g_n) \quad \text{and} \quad \boxed{\ } = 0 = $$

for $1 \leq i \leq n-1$.

Let $\Delta_n = \;$ denote the bracket evaluation of this clo-

sure. Then must be a multiple of by the argument

of Proposition 1. But $\; = x \;$ imples that $\; =$

$x \;$, hence $\Delta_{n+1} = x\Delta_n$ and $x = \Delta_{n+1}/\Delta_n$:

$$= \frac{\Delta_{n+1}}{\Delta_n} \qquad .$$

Now, if $\; = \; + y \;$ it follows that

(by composing with U_{n+1})

$$0 = \quad \vcenter{\hbox{[diagram]}} \quad + y \quad \vcenter{\hbox{[diagram]}} \quad .$$

Hence

$$0 \;=\; \vcenter{\hbox{[diagram]}} \;+\; y\, \frac{\Delta_{n+1}}{\Delta_n} \;\vcenter{\hbox{[diagram]}} \;=\;$$

$$=\; \vcenter{\hbox{[diagram]}} \;+\; y\, \frac{\Delta_{n+1}}{\Delta_n} \;\vcenter{\hbox{[diagram]}}$$

since it is easy to see that

$$\vcenter{\hbox{[diagram]}} \;=\; \vcenter{\hbox{[diagram]}} \quad .$$

Therefore $y = -\Delta_n/\Delta_{n+1}$ and the recursion takes the form

From this it follows that $\Delta_{n+2} = \Delta_{n+1}d - \Delta_n$ with $\Delta_0 = 1, \Delta_{-1} = 0$ (by taking thes strand closure.)

These remarks show that the coefficients in the recursion are determined by the requirements that g_n be a projector. The same considerations give the inductive proof of the existence of these projectors.

Chapter 4

The 3-Vertex

4.1 A Special Sum of Tangles

In the next contruction we associate a sum of tangles with a 3-valent graphical vertex. The vertex and the sum of tangles will then be used interchangeably. Each line on the vertex is labelled with a positive integer a, b or c as shown below.

It is assumed that $(a+b-c), (a+c-b)$ and $(b+c-a)$ are each a positive even integer. Then, letting

$$
\begin{array}{rcl}
i & = & (a+b-c)/2 \\
j & = & (a+c-b)/2 \\
k & = & (b+c-a)/2,
\end{array}
$$

the 3-vertex is defined by the equation.

Definition 3

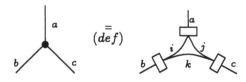

Here it is understood that each Temperley-Lieb projector is to be expanded fully, so that the whole vertex is a sum of tangles that are free from under-or-over crossings. For example,

Lemma 2 *With $d = -A^2 - A^{-2}$,*

Proof: $d = -A^2 - A^{-2}$. Apply this expansion three times to the given 3-vertex:

4.2   A Fundamental Twist

The following formula instantiates one of the motivations for this definition of 3-vertex. The control over this twist given by the formula is central in obtaining the 3-manifold invariants of Chapter 12.

Proposition 3 (Twist Formula)

$$\tilde{T} = \quad = (-1)^{\frac{b+c-a}{2}}$$

$$A^{\frac{[b(b+2)+c(c+2)-a(a+2)]}{2}}$$

Proof: First consider the result of twisting a tangle of the form

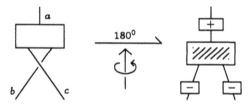

by 180°, keeping the end-points fixed. Here a, b, c denote the number of parallel strands in these input and output cables. the result is to turn the tangle by 180° and to put 180° twists in the input and output cables of the tangles, as shown. We have denoted 180° cable twists by boxes labelled $\boxed{+}$ and $\boxed{-}$. Thus

$$\boxed{+} = \quad \text{and} \quad \boxed{-} = \quad . \text{ Also,}$$

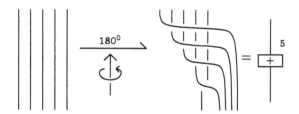

Figure 5: Form of the 180° Twist

Now suppose that the tangle in this picture is a planar tangle, free from any internal over or under-crossings, and that it has no free loops. Then it has the form

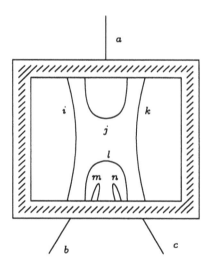

where $i + 2j + k = a, i + l + 2m = b, k + l + 2n = c$.

Here i denotes the number, $\lambda(a, b)$, of lines running from a to b. Also let $k = \lambda(a, c)$, $j = \lambda(a, a)$, $l = \lambda(b, c)$, $m = \lambda(b, b)$ and $n = \lambda(c, c)$.

Thus in the form 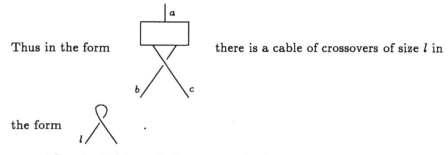 there is a cable of crossovers of size l in

the form 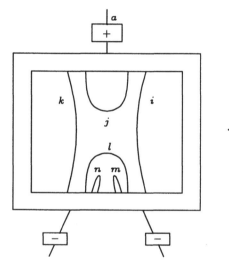 .

After the twist, we find crossovers in the pattern:

Individually, this means the following pick-up of crossovers:

and a loss of the crossover .

Now we need two facts that are easy to verify, and whose details are left to the reader (but compare with Figures 5,6):

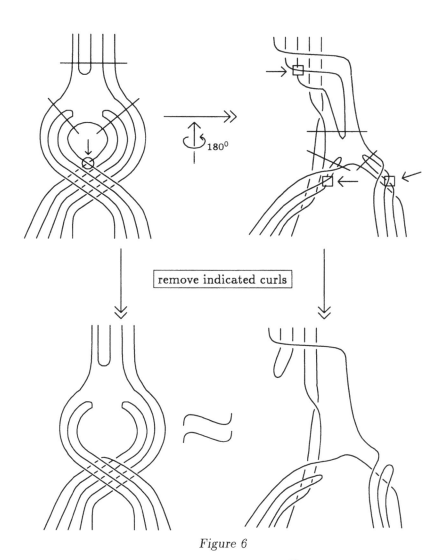

Figure 6

(i) The original tangle with l curls of type ⧖ removed (i.e. replace ⧖ by ∩.) is regularly isotopic to the rotated tangle with $j+n+m$

curls removed: j of type ⋎ , $n+m$ of type ⋌ .

Hence if T and T' denote the original tangle and the rotated tangle, respectively, then $(-A^3)^{-l}T = (-A^3)^{-j}(-A^{-3})^{-n-m}T'$ in the sense of bracket evaluations. Hence

$$T = (-A^3)^{l+n+m-j}T'$$

(ii) The sum $l + n + m - j$ is *independent* of the particular choice of connections in the flat tangle. Hence it may be calculated from the special case shown below:

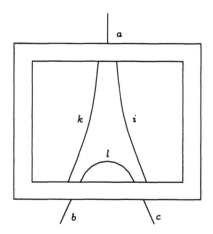

In the special case, $n = m = j = 0$ and $l = (b + c - a)/2$.

Hence, we conclude that $\tilde{T} = (-A^3)^{\frac{b+c-a}{2}}\tilde{T}'$ where $\tilde{T}' =$.

We can make this conclusion because $\tilde{T}$ (the twisted 3-vertex) is a sum of tangles of the sort considered in the argument as given up to this point. Since the coefficient $(-A^3)^{(b+c-a)/2}$ is common to all these tangles, we can write a global formula above.

Finally, we note that

$$\begin{array}{c} \boxed{+} \\ a \\ \Box \end{array} = A^{-(a-1)a/2} \begin{array}{c} a \\ \Box \end{array}$$

$$\begin{array}{c} \boxed{-} \\ a \\ \Box \end{array} = A^{(a-1)a/2} \begin{array}{c} a \\ \Box \end{array}$$

since the 180° twist braid on a strands has $(a-1)a/2$ crossings, and the Temperley-Lieb projector only allows one state of the braid to survive (the identity state, consisting, for $\boxed{-}$, in all factors of A and for $\boxed{+}$ in all factors of A^{-1}). Thus $\tilde{T}' = A^{[-(a-1)a+(b-1)b+(c-1)c]/2}\tilde{T}''$, where

$$\tilde{T}'' = \overbrace{}^{a} \quad . \text{ Hence}$$

$$\begin{aligned} \tilde{T} &= (-A^3)^{\frac{-a+b+c}{2}} (A)^{\frac{-(a-1)a+(b-1)b+(c-1)c}{2}} \tilde{T}'' \\ &= (-1)^{\frac{b+c-a}{2}} A^{[-a(a+2)+b(b+2)+c(c+2)]/2}\tilde{T}''. \end{aligned}$$

This completes the proof. ∎

As an example of previous Proposition we get

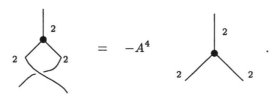

4.3   Invariants of Trivalent Embedded Graphs: $P_a(G)$

The main force of Proposition 3 is, at this stage, that it allows us to construct invariants of embedded graphs with trivalent nodes. (Later, we will use the same formalism to construct invariants of 3-manifolds.) There are various schemes for putting this into practice. The simplest is to define $P_a(G)$ for such a graph G, represented as a diagram with over and under-passes, i.e. as a link diagram containing trivalent nodes. Then $P_a(G) = < a * G >$ where $<\ >$ denotes a bracket polynomial evaluation, and $a*G$ is the network obtained by replacing each strand of G by a parallel strands, and each vertex of G by a 3-vertex of the form

This is the simplest labelling scheme. Proposition 3 allows us to conclude that $P_a(G)$ is, up to sign and multiplicative factors that are a power of A, an invariant of the *framed isotopy class* of the graph embedding G. Framed isotopy is defined as follows.

Definition 4 *Two trivalent graph diagrams are framed isotopic if one can be obtained from the other by a combination of the following operations:*

(1) regular isotopy (i.e. Reidemeister moves II and III) on parts of diagram free from graphical vertices.

(2) the following graph moves

(a)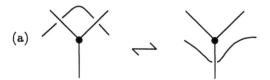

(b)

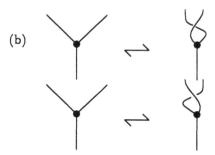

Remark 5 *A quasi-physical model of framed isotopy ensues if we replace the lines of the graph by ribbons, so that a curl* ⌒ *is modelled by a twist in a ribbon* ⌒ *. In any case, we have proved*

Proposition 4 *Let $P_a(G)$ be defined as above, $P_a(G) = < a * G >$. Then $P_a(G)$ is an invariant of framed isotopy of G, up to multiples by powers of $(-1)^a A^{a(a+2)/2}$.*

Proof: This follows directly from the twist formula and the regular isotopy invariance of the bracket. ■

The formula $P_a \left(\vphantom{\bigg|} \right) = P_a \left(\vphantom{\bigg|} \right)$ follows from invariance of the bracket under type II and type III moves, since the 3-vertex is expanded as a sum of planar tangles.

4.4 The Case of $P_2(G)$

In the case of $P_2(G)$ it is worth recording a list of properties that allow its recursive computation. To this end, a shorthand useful. Let $P_2(G)$ be denoted by $[G]$. Thus, for a link K, $[K] = < 2 * K >$ and $\left[\vphantom{\bigg|} \right]$ will

mean $\left\langle \; 2 \quad 2 \; \right\rangle = \left\langle \; \right\rangle$.
$\qquad\qquad\qquad 2$

We can carry this abbreviation further with the convention

$$[\longrightarrow\!\bullet] = < \supset > \quad .$$

Thus $[\,\bullet\!\!-\!\!\!-\!\!\bullet\,] = < \bigcirc \!\!\! > = d$ while $\left[\,\bigcirc\,\right] = \Big\langle\!\Big\langle \bigcirc \Big\rangle\!\Big\rangle = d^2$

where $d = -A^2 - A^{-2}$.

Applying this formalism to 2, we have

$$\left[\;\bigtriangleup\;\right] = \left[\;\bigtriangleup\;\right]$$

$$-\frac{1}{d}\left(\left[\;\bigtriangleup\;\right] + \left[\;\bigtriangleup\;\right] + \left[\;\bigtriangleup\;\right]\right) + \frac{2}{d^2}\left[\;\bigtriangleup\;\right]$$

where $\left[\;\bigtriangleup\;\right] = \Big\langle\;\bigtriangleup\;\Big\rangle$ by definition. For example,

Lemma 3 $P_2\left(\bigcirc\!\!\!\!\!\bigcirc\right) = \left[\bigcirc\!\!\!\!\!\bigcirc\right] = d^3 - 3d + 2d^{-1}$ *where* $d = -A^2 - A^{-2}$.

Proof:

$$\left[\bigcirc\!\!\!\!\!\bigcirc\right] = \left[\bigcirc\!\!\!\!\!\bigcirc\right] - \frac{1}{d}\left[\bigcirc\!\!\!\!\!\bigcirc\right] - \frac{1}{d}\left[\bigcirc\!\!\!\!\!\bigcirc\right]$$

$$-\frac{1}{d}\left[\bigcirc\!\!\!\!\!\bigcirc\right] + \frac{2}{d^2}\left[\bigcirc\!\!\!\!\!\bigcirc\right]$$

But note that

$$\bigtriangleup = \bigtriangleup = 0$$

since the Temperley-Lieb projector annihilates the loop. Therefore,

$$\left[\bigoplus\right] = \left[\bigoplus\right] \because \left[\bigoplus\right] = \left[\bigoplus\right]$$

$$-\frac{1}{d}\left[\bigoplus\right] - \frac{1}{d}\left[\bigoplus\right] - \frac{1}{d}\left[\bigoplus\right] + \frac{2}{d^2}\left[\bigoplus\right]$$

$$= \left\langle\bigoplus\right\rangle - \frac{1}{d}\left\langle\bigoplus\right\rangle - \frac{1}{d}\left\langle\bigoplus\right\rangle - \frac{1}{d}\left\langle\bigoplus\right\rangle$$

$$+ \frac{2}{d^2}\left\langle\bigoplus\right\rangle \qquad \left[\bigoplus\right] = d^3 - 3d + 2d^{-1}.$$

The following formulas [Kau85] complete the algorithm for computing $P_2(G)$. They are easily proved by using the bracket expansion for double-stranded links. ∎

Proposition 5 *Let $[G]$ denote $P_2(G)$ with the diagrammatic conventions described above. Then the following formulas give an algorithm for computing $[G]$.*

$$(i) \quad \left[\bigtimes\right] = A^4\left[\bigasymp\right] + A^{-4}\left[\bigsupset\bigsubset\right]$$

$$+ A^2\left[\bigasymp\right] + A^2\left[\bigasymp\right]$$

$$+ A^{-2}\left[\bigsupset\bigsubset\right] + A^{-2}\left[\bigsupset\bigsubset\right]$$

$$+ (A^2 + A^{-2})\left[\bigtimes\right]$$

$$+ \left[\bigtimes\right] + \left[\bigtimes\right] + \left[\bigtimes\right]$$

$$+ \left[\bigtimes\right]$$

Here $\left[\,\times\!\!\!\times\,\right] = \left\langle\,\asymp\,\right\rangle$ by definition.

(ii) $\left[\,\begin{smallmatrix}\mathcal{R}\\\wedge\end{smallmatrix}\,\right] = A^{-8}\,[\,\frown\,]+(A^{-6}-A^{-2})\,[\,-\!\!\bullet\quad\bullet\!\!-\,]$

$\left[\,\begin{smallmatrix}\mathcal{R}\\\wedge\end{smallmatrix}\,\right] = A^{8}\,[\,\frown\,]+(A^{6}-A^{2})\,[\,-\!\!\bullet\quad\bullet\!\!-\,]$

(iii) $[\,\bullet\!\!-\!\!-\!\!\bullet\,] = d$

$\left[\,\bigcirc\,\right] = d^2$

$\left[\,\stackrel{\curvearrowright}{\bullet\quad\bullet}\,{}^{G}\,\right] = d[G]$

$\left[\,\bigcirc\,{}^{G}\,\right] = d^2[G]$

(iv) $\left[\,\begin{smallmatrix}\bigwedge\\\bullet\end{smallmatrix}\,\right] = \left[\,\begin{smallmatrix}\bigwedge\\\circ\end{smallmatrix}\,\right] - \frac{1}{d}\left[\,\begin{smallmatrix}\big|\bigwedge\\\bullet\end{smallmatrix}\,\right]$

$-\frac{1}{d}\left[\,\begin{smallmatrix}\bigwedge\\\bullet\end{smallmatrix}\,\right] - \frac{1}{d}\left[\,\begin{smallmatrix}\bullet\\\cup\end{smallmatrix}\,\right] + \frac{2}{d^2}\left[\,\begin{smallmatrix}\big|\\\bullet\,\bullet\end{smallmatrix}\,\right]$

Proof: (i) and (ii) follow from the bracket expansion, $\left[\,\frown\!\!\smile\,\right] =$ $\left\langle\,\frown\!\!\smile\,\right\rangle$, by a calculation that we here omit. (iii) and (iv) have been discussed just prior to this proof.   ∎

Remark 6 *An interesting (and simpler) invariant that is related to $P_2(G)$ is the invariant of regular isotopy that corresponds to replacing each component of the link L by two parallel planar strands with a projector in the cable. In the notation of Proposition 5 this invariant has the following* *recursion formulas:*

$$[\,\times\,]_2 \;=\; A^4[\,\smile\,]_2 + A^{-4}[\,\supset\subset\,]_2 + (A^2 + A^{-2})[\,\times\!\!\times\,]_2$$

$$[\,\text{⚬}\,]_2 \;=\; A^{-8}[\,\frown\,]_2$$

$$[\,\text{⚬}\,]_2 \;=\; A^{8}[\,\frown\,]_2$$

$$[\,\bigcirc\,]_2 \;=\; \Delta_2(d) = d^2 - 1, \; d = -A^2 - A^{-2}$$

$$[\,\times\!\!\times\,]_2 \;=\; \Big\langle\,\text{⧓}\,\Big\rangle$$

$$[\,\frown\!\!\frown\,]_2 \;=\; \langle\,\text{⊨}\,\rangle \quad .$$

The last two formulas define the relationship of this invariant with the standard bracket polynomial. We denote this invariant by $[K]_2$. Note that the state summation indicated by these formulas writes $[K]_2$ as a sum of 4-valent plane graph evaluations that themselves involve expanding networks of projectors. Note that

$$[\,\times\,]_2 - [\,\times\,]_2 = (A^4 - A^{-4})\left([\,\smile\,]_2 - [\,\supset\subset\,]_2\right)$$

with

$$[\,\text{⚬}\,]_2 \;=\; A^{-8}[\,\frown\,]_2$$

$$[\,\text{⚬}\,]_2 \;=\; A^{8}[\,\frown\,]_2 \quad .$$

This shows that the normalized (ambient isotopy) version of $[K]_2$ is a special case of the Dubrovnik version of the Kauffman polynomial [Kau90b].

Chapter 5

Properties of Projectors and 3-Vertices

5.1 Vanishing Conditions

In this section basic facts about 3-vertices will be established for $\sqrt{q} = \mathcal{A}$ a root of unity. We then use these methods to determine conditions when the evaluations of a network of the form  is non-zero.

Notation: A network and its evaluation will be used interchangebly. Thus

 will denote the bracket evaluation of the network summation

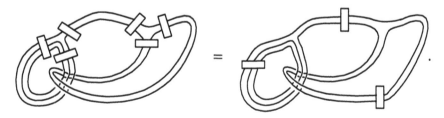

If we wish to emphasize a particular value of q, or to emphasize or parenthesize the evaluation, then we shall enclose the network in a square bracket with subscript q:

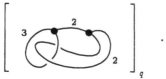

Recall from Section 2 that $f_{n-1} = \boxed{}^{\,n}$ with $f_{n-1}^2 = f_{n-1}$, $f_{n-1}U_i = 0$, $1 \leq$

$n - 1$. Furthermore, $tr(f_{n-1}) = ^{n}\!\bigcirc\!\boxed{} = \Delta_n$ where

$$
\begin{aligned}
\Delta_n &= [(-q)^{n+1} - (-q)^{-n-1}]/((-q) - (-q^{-1})) \\
&= (-1)^n[q^{n+1} - q^{-n-1}]/(q - q^{-1}).
\end{aligned}
$$

$(\sqrt{q} = A)$.

If q is a $2r$-th primitive root of unity, then $q^r = -1$, $q = e^{i\pi/r}$. Thus

$$
\Delta_{r-1} = (-1)^{r-1}[-1 + 1]/(q - q^{-1}),
$$

hence $\Delta_{r-1} = 0$. On the other hand, if $0 \leq l \leq r - 2$, then

$$
\begin{aligned}
\frac{q^{l+1} - q^{l-1}}{q - q^{-1}} &= \frac{e^{i\pi(l+1)/r} - e^{-i\pi(l+1)/r}}{e^{i\pi/r} - e^{-i\pi/r}} \\
&= \frac{\sin(\pi(l+1)/r)}{\sin(\pi/r)} \neq 0.
\end{aligned}
$$

Thus, we have shown

Lemma 4 *Let* $q = e^{i\pi/r}$. *Let* $\Delta_n = (-1)^n[q^{n+1} - q^{-n-1}]/(q - q^{-1})$ *as above. Then* $\Delta_{r-1} = 0$ *while* $\Delta_l \neq 0$ *for* $0 \leq l \leq r - 2$.

Proof: The proof consists in the discussion preceding the Lemma. ∎

Comment: From Lemma 4 and the remarks preceding it we get for generic q,

$$
^{n}\!\bigcirc\!\boxed{} = \Delta_n
$$

and therefore, at $q = e^{i\pi/r}$,

$$\overset{l}{\ominus} \neq 0 \quad \text{for} \quad 0 \le l \le r-2$$

while

$$\overset{r-1}{\ominus} = 0.$$

Lemma 5 *In fact, at $q = e^{i\pi/r}$, we have the identity* $\boxed{\overset{r-1}{}} = 0$.

Proof: Regard this projector as part of a larger closed network. Consider the expansion of all projectors in the network except the projector explicity shown. The outside connection of any term that survives must be straight. Otherwise there are turn-backs. Thus there exists a λ such that

$$\left[\boxed{\overset{r-1}{}} \right] = \lambda \left[\overset{r-1}{\boxed{}\ominus} \right] = \lambda \Delta_{r-1} = 0. \qquad \blacksquare$$

Example For $r = 3, d = -q - q^{-1} = -1$, since $q = e^{i\pi/r}$. Thus

$$0 = \boxed{} = \Big) \Big(-\frac{1}{d} \overset{\cup}{\cap} = \Big) \Big(+\overset{\cup}{\cap}.$$

Thus at $r = 3$,

$$\Big) \Big(= -\overset{\cup}{\cap}.$$

Remark 7 *It is useful to point out the relationship between the Chebyschev polynomials and the coefficients μ_n in the recursion formula for projectors in the Temperley-Lieb algebra. Recall from Definition 1 that we have $\mu_1 = d^{-1}, \mu_{k+1} = (d - \mu_k)^{-1}$ where d is the loop value in the Temperley-Lieb algebra. If we define Δ_n for $n = 0, 1, 2, \ldots$ by $\Delta_0 = 1, \Delta_1 = d, \Delta_{n+1} = d\Delta_n - \Delta_{n-1}$, then it follows at once that $\mu_n = \Delta_{n-1}/\Delta_n$. On the other* *hand, if Δ_n is defined in this way and $d = x + x^{-1}$, then it is easy to prove that $\Delta_n = (x^{n+1} - x^{-n-1})/(x - x^{-1})$.*

Proof: $\Delta_{n+1} = (x + x^{-1})\Delta_n - \Delta_{n-1}$

$$\Rightarrow \Delta_{n+1} - x^{-1}\Delta_n = x\Delta_n - \Delta_{n-1} = x(\Delta_n - x^{-1}\Delta_{n-1}).$$

Hence $\Delta_n - x^{-1}\Delta_{n-1} = x^{n-1}(\Delta - x^{-1}) = x^n$. Similarly, $\Delta_n - x\Delta_{n-1} = x^{-n}$. Thus $(x - x^{-1})\Delta_{n-1} = x^n - x^{-n}$. Whence

$$\Delta_n = (x^{n+1} - x^{-n-1})/(x - x^{-1}). \qquad \blacksquare$$

Throughout this Chapter, unless stated otherwise, we shall assume that $\Delta_n = \Delta_n(x = -e^{i\pi/r})$. In other words, we take $x = -A^2$, $d = -A^2 - A^{-2}$ and specialize at $q = e^{i\pi/r}$ where $\sqrt{q} = A$.

For the case of generic d, a few more remarks about the projectors are in order.

For example,

Note that the coefficients of left-right symmetric terms are equal. This is easily proved via the braiding approach of Definition 2, and we leave it as an exercise for the reader. We have used repeatedly, and will use again, the

fact that $= 1_n + \mathcal{U}_n$ where $\mathcal{U}_n$ is a sum of Temperley-Lieb elements,

each containing a turn-back.

Lemma 6

Proof: The only term which survives in the expansion of the bottom projector is the "straight ahead" term. All the other terms have turn-backs, and are null.   ∎

Lemma 7 *The network*   *is null if $a \neq b$. If $a = b$, then*

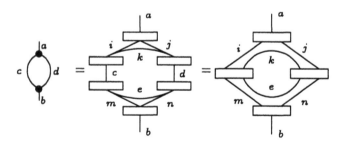

This result holds for Temperley-Lieb algebra at any loop value where $\Delta_a \neq 0$.

Proof: Assume that $a > b$. Clearly a and b have the same parity, otherwise the network is null. By writing out the definition of the 3-vertices, we have

where

$$i = \frac{a+c-d}{2} \qquad l = \frac{c+d-b}{2}$$
$$j = \frac{a+d-c}{2} \qquad m = \frac{c+b-d}{2}$$
$$k = \frac{c+d-a}{2} \qquad n = \frac{b+d-c}{2}.$$

Thus $i > m, j > n, k < l$. Rewriting, we find with $e = (a - b)/2$

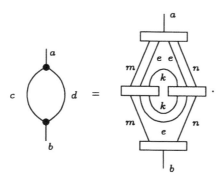

Because $e > 0$, it follows that each term in the expansion of the middle two projectors, will contain a turn-back with respect to the a-projector above. Hence the network is null.

Now assume that $b = a$. Consider,

$$(x + y = a)$$

Hence 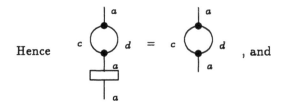 , and

since only straight-ahead terms survive the extra projector. Thus

Hence . Since λ is determined if

$\Delta_a \neq 0$, this completes the proof of the Lemma.   ∎

5.2   Interaction with Curls and Loops

We end this Chapter with two basic propositions about the interaction of a projector with curls and loops.

Proposition 6

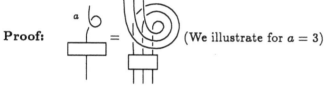

Proof:

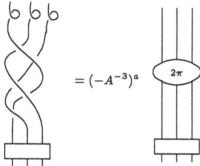 (We illustrate for $a = 3$)

$$= (-A^{-3})^a \quad \boxed{2\pi}$$

$$= (-A^{-3})^a (A^{-1})^{a(a-1)}$$

(There are $a(a-1)$ crossings in the 2π twisted cable.) Thus

$$\overset{a}{\underset{}{\boxed{}}} = (-1)^a A^{-a(a+2)} \overset{a}{\underset{}{\boxed{}}} .$$

This proves (i), and (ii) proceeds in the same way. ∎

Proposition 7 *([Lic92]).*

$$= (-A^{2m+2} - A^{-2m-2})$$

Proof: Again, we illustrate with $m = 3$.

$$= A \qquad + A^{-1}$$

$$= A^m \qquad + A^{-m}$$

(Keep using the fact that projectors annihilate turn-backs.)

$$= A^m(-A^3) \qquad + A^{-m}(-A^{-3})$$

$$= (-A^{m+3})A^{m-1} \qquad + (-A^{-m-3})A^{-m+1}$$

$$= (-A^{2m+2} - A^{-2m-2})$$

These identities will be particularly useful in Chapter 12.

Chapter 6

θ-Evaluations

We now turn to the problem of evaluating *theta-nets*. A theta-net evaluation is denoted $\theta(a, b, c)$. This denotes the value in the Temperley-Lieb algebra of the net .

For the rest of this Chapter we take $q = e^{i\pi/r}$, $d = -q - q^{-1} = -2cos(\pi/r)$. (This d is the loop value in the Temperley-Lieb algebra.) Thus, by Proposition 4, $\Delta_{r-1} = 0$ while $\Delta_l \neq 0$ for $0 \leq l \leq r - 2$.

6.1 Recursive Relations

The first Sections record our initial investigation of the θ-net. We produce a recursive algorithm that leads directly to an explicit evaluation. If one knows the answer, then shorter inductive proofs are available. This issue is addressed in Chapter 8 where we show how the special case of $A = -1$ provides the key to the formulas for both the θ and tetrahedral nets. (This special case is related to the classical theory of quantized angular momentum.)

The reader who is primarily interested in the formula for the θ-net should skip the technicalities of this chapter and simply read the statement of Corollary 2 and the comments that follow it at the end of this Chapter. The first simplification is as follows.

Lemma 8 $\theta(a, b, c) =$

where $m = \dfrac{a+b-c}{2}$, $n = \dfrac{b+c-a}{2}$, $p = \dfrac{a+c-b}{2}$.

Proof:

$$\theta(a, b, c) \quad = $$

$$=$$

■

Definition 5 $Net(m, n, p) =$

From now on, all our work will focus on the evaluation of $Net(m, n, p)$. In the case $p = 0$, we get

$$Net(m, n, 0) \quad = $$

$$= \Delta_{m+n}$$

The case $p = 1$ is simple to deal with.

Lemma 9 $Net(m, n, 1) = (d - \mu_{mn} - \mu_n)\Delta_{m+n}$. *Here* $d = -q - q^{-1} = -2cos(\pi/r)$ *is the loop value, and the* μ_k *are defined via Definition 1.*

Proof: Recall the basic recursion relation for projectors

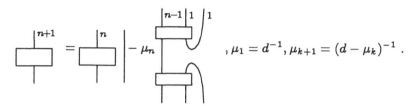

$$\mu_1 = d^{-1}, \mu_{k+1} = (d - \mu_k)^{-1} .$$

Apply this to $Net(m, n, 1)$ as shown below.

The last network is equivalent to 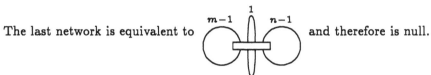 and therefore is null.

The first network is directly $d\Delta_{m+n} = d\ Net(m, n, 0)$ by the calculation of $Net(m, n, 0)$ prior to this Lemma. The second and third nets are each equivalent to $Net(m, n, 0)$. Thus $Net(m, n, 1) = d\Delta_{n+m} - \mu_m\Delta_{n+m} - \mu_n\Delta_{n+m}$. This completes the proof. ∎

Definition 6 *Let $Net(m, n, p_e, p_i)$, for $p_e + p_i = p - 1 \geq 1$ denote the network evaluation indicated below.*

$$Net(m, n, p_e, p_i) =$$

The next lemma will make the reason for this definition apparent.

Lemma 10 *For $p \geq 2$*

$$Net(m, n, p) = (d - \mu_{m+p-1} - \mu_{n+p-1})Net(m, n, p - 1)$$
$$+ \mu_{m+p-1}\mu_{n+p-1}Net(m, n, 1, p - 2).$$

Proof:

$$Net(m, n, p) =$$

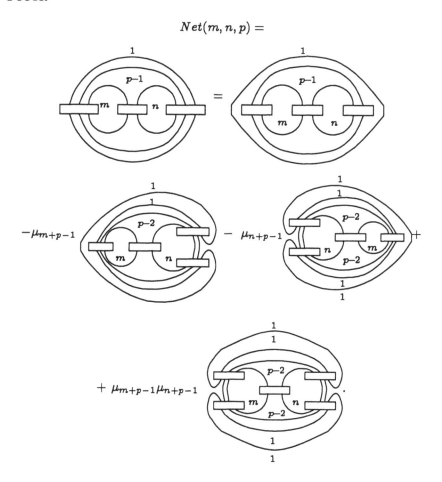

The last network is equivalent to $Net(m, n, 1, p - 2)$. (In fact $Net(m, n, 1, p-2)$ is simply a re-configuration of this last net.) This completes the proof of the lemma. ∎

The next result will express $Net(m, n, p_e, p_i)$ in terms of $Net(m, n, p_e + 1, p_i - 1)$.

Lemma 11

$$(a) \quad Net(m, n, p-1, 0) \;=\; (d - \mu_m - \mu_n)Net(m, n, p-1)$$

$$(b) \quad Net(m, n, pe, p_i) \;=\; (d - \mu_{m+p_i} - \mu_{n+p_i})Net(m, n, p-1)$$

$$+\mu_{m+p_i}\mu_{n+p_i}Net(m, n, pe+1, p_i-1).$$

Proof:

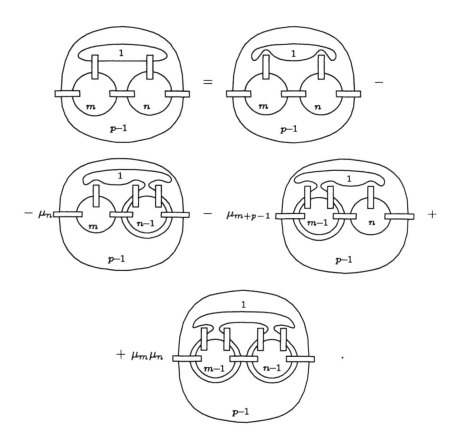

A moment's inspection reveals that the last network is null. Thus part (a)

follows at once. The next expansion proves part (b).

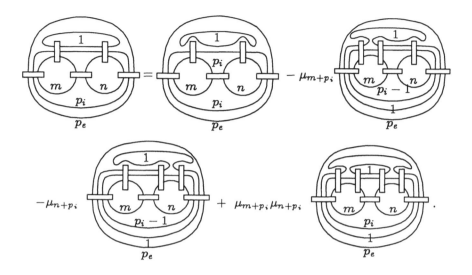

This completes the proof. ∎

6.2   Shaping the Recursion

We are ready to prove the basic recursion to get the θ-symbols.

For fixed m and n, let $\alpha_j = d\Delta_{m+j}\Delta_{n+j} - \Delta_{m+j-1}\Delta_{n+j} - \Delta_{m+j}\Delta_{n+j-1}$. Let $\beta_j = \Delta_{m+j}\Delta_{n+j}$.

[Note that me take the loop value to be $d=-2cos(\pi/r)$, and $\Delta_n=(x^{n+1}-x^{-n-1})/(x-x^{-1})$ with $x = -e^{i\pi/r}$, and $\mu_m = \Delta_{m-1}/\Delta_m$.]

By judicious application of Lemma 11 we shall find a factor, $\rho(m, n, p)$, such that $Net(m, n, p) = \rho(m, n, p)Net(m, n, p-1)$. The shape of this recursion is indicated in Figure 7.

Lemma 12 *Assuming that $m+n+p \leq r-1$, there exists a factor $\rho(m,n,p)$ such that $Net(m,n,p) = \rho(m,n,p)Net(m,n,p-1)$. The factor is given by the following formula:*

$$\rho(m,n,p) = \frac{1}{\beta_{p-1}} \sum_{j=0}^{p-1} \alpha_j.$$

Upon summation, one obtains the more compact version

$$\rho(m,n,p) = d - \mu_{m+p-1} - \frac{\Delta_n \Delta_{n-1}}{\Delta_{m+p-1}\Delta_{n+p-1}}.$$

Proof: Recall the formulas of Lemma 11:

(a) $Net(m,n,p-1,0) = (d - \mu_m - \mu_n)Net(m,n,p-1)$

(b) $Net(m,n,pe,p_i) = (d - \mu_{m+p_i} - \mu_{n+p_i})Net(m,n,p-1)$

$\qquad\qquad\qquad\qquad + \mu_{m+p_i}\mu_{n+p_i}Net(m,n,pe+1,p_i-1).$

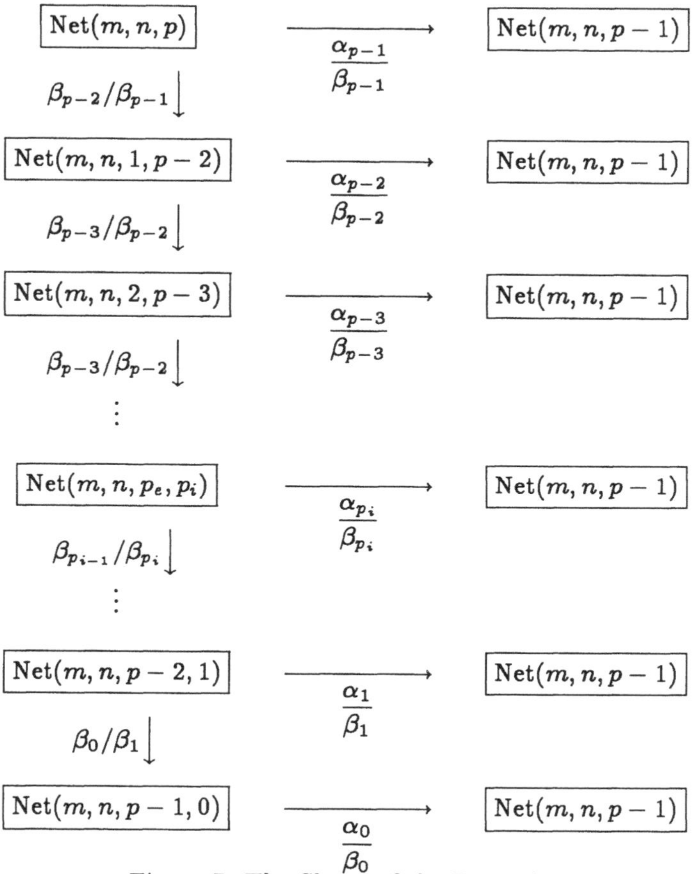

Figure 7: The Shape of the Recursion

Since $\mu_m = \Delta_{m-1}/\Delta_m$, it follows that

$$d - \mu_m - \mu_n = \frac{d\Delta_m\Delta_n - \Delta_{m-1}\Delta_n - \Delta_m\Delta_{n-1}}{\Delta_m\Delta_n} \quad .$$

Thus we can re-write (a) and (b) as follows.

(a) $Net(m, n, p-1, 0) \;=\; (\alpha_0/\beta_0)Net(m, n, p-1)$

(b) $Net(m, n, pe, p_i) \;=\; (\alpha_{p_i}/\beta_{p_i})Net(m, n, p-1)$

$$(\beta_{p_i-1}/\beta_{p_i})Net(m, n, pe+1, p_i-1).$$

This re-write will serve our calculation, and constitues the shape of the recursion shown in Figure 7.

The factor $\rho(m, n, p)$ is a sum $\sum_{j=0}^{p-1} \tau(m, n, j)$ where each term $\tau(m, n, j)$ is in $1-1$ correspondence with the pendent vertex at height j in the recursion tree of Figure 7. Its value is the product of the weights along the path to the root:

$$
\begin{aligned}
\tau(m, n, p-1) &= \alpha_{p-1}/\beta_{p-1} \\
\tau(m, n, p-2) &= (\alpha_{p-2}/\beta_{p-2})(\beta_{p-2}/\beta_{p-1}) = \alpha_{p-2}/\beta_{p-1} \\
&\vdots \\
\tau(m, n, p_i) &= (\alpha_{p_i}/\beta_{p_i})(\beta_{p_i}/\beta_{p_i-1})\cdots(\beta_{p-2}/\beta_{p-1}) = \frac{\alpha_{p_i}}{\beta_{p-1}} \\
&\vdots \\
\tau(m, n, 1) &= (\alpha_1/\beta_1)(\beta_1/\beta_2)\cdots(\beta_{p-2}/\beta_{p-1}) = \alpha_1/\beta_{p-1} \\
\tau(m, n, 0) &= (\alpha_0/\beta_0)(\beta_0/\beta_1)\cdots(\beta_{p-2}/\beta_{p-1}) = \alpha_0/\beta_{p-1}.
\end{aligned}
$$

The $\beta_j' s$ are indeed non-null. For $\beta_j = \Delta_{m+j}\Delta_{n+j}$ to be null we must have either $m + j = r - 1$ or $n + j = r - 1$. But we are assuming $1 \le p, j \le p$ and $m + n + p \le r - 1$. Thus the above coefficients make sense, and we get

$$
\rho(m, n, p) = \frac{1}{\beta_{p-1}} \sum_{j=0}^{p-1} \alpha_j.
$$

Now, computing, we find

$$
\begin{aligned}
\beta_{p-1}\rho(m, n, p) =\ &d\Delta_{m+p-1}\Delta_{n+p-1} - \Delta_{m+p-2}\Delta_{n+p-1} - \Delta_{m+p-1}\Delta_{n+p-2} \\
&+d\Delta_{m+p-2}\Delta_{n+p-2} - \Delta_{m+p-3}\Delta_{n+p-2} - \Delta_{m+p-2}\Delta_{n+p-3} \\
&+ \cdots \\
&+d\Delta_{m+p_i}\Delta_{n+p_i} - \Delta_{m+p_i-1}\Delta_{n+p_i} - \Delta_{m+p_i}\Delta_{n+p_i-1} \\
&+ \cdots \\
&+d\Delta_{m+1}\Delta_{n+1} - \Delta_m\Delta_{n+1} - \Delta_{m+1}\Delta_n \\
&+d\Delta_m\Delta_n - \Delta_{m-1}\Delta_n - \Delta_m\Delta_{n-1} \\
=\ &d\Delta_{m+p-1}\Delta_{n+p-1} - \Delta_{n+p-2}\Delta_{n+p-1} \\
&+ \sum_{j=0}^{p-2}(d\Delta_{m+j} - \Delta_{m+j+1} - \Delta_{m+j-1})\Delta_{n+j} - \Delta_m\Delta_{n-1}.
\end{aligned}
$$

By the defining recursion for the $\Delta_n' s$, each term between round brackets in the last summation is zero. We obtain the formula

$$
\rho(m, n, p) = d - \mu_{m+p-1} - \frac{\Delta_m\Delta_{n-1}}{\Delta_{m+p-1}\Delta_{n+p-1}}.
$$

This completes the proof.   ∎

6.3 A Formula for the θ-nets

From the above explicit value for $\rho(m, n, p)$ we get

Proposition 8 *Assume that*

$$\left\{ \begin{array}{l} a, b, c \leq r - 1 \\ m = \frac{a+b-c}{2}, \quad n = \frac{b+c-a}{2}, \quad p = \frac{a+c-b}{2} \\ m + n + p \leq r - 1 \end{array} \right\}.$$

Then $\rho(m, n, p) = 0$ *if and only if* $m + n + p = r - 1$. *Equivalently,* $\theta(a, b, c) = 0$ *if and only if* $a + b + c = 2r - 2$.

Proof: By re-writing ρ in terms of $x = -q$, we get

$$\begin{aligned} \rho(m, n, p) &= (x + x^{-1}) - \frac{x^{m+p-1} - x^{-m-p+1}}{x^{m+p} - x^{-m-p}} \\ &\quad - \left(\frac{x^{m+p} - x^{-m-1}}{x^{m+p} - x^{-m-p}} \right) \left(\frac{x^n - x^{-n}}{x^{n+p} - x^{-n-p}} \right). \end{aligned}$$

In terms of x, the equality $\rho(m, n, p) = 0$ is equivalent to (after expansion and cancellations)

$$x^{2(m+n+p+1)} = 1.$$

But x is a primitive $2r$-root of unity. Therefore the equality holds if and only if $2(m + n + p + 1) = 2r$. That is, $m + n + p = r - 1$.

To prove the result for $\theta(a, b, c)$ it is enough to treat the case $p = 0$. But $\Delta_{m+n} = \Delta_{r-1} = 0$.

Corollary 1

$$\theta(a, b, c) = \left[\begin{array}{c} \\ a \bigodot^b c \\ \\ \end{array} \right] = \left[\begin{array}{c} p \\ \bowtie \\ m \quad n \end{array} \right]$$

$$= \left(\prod_{j=1}^{p} \rho(m, n, j) \right) \Delta_{m+n}$$

for $m + n + p \leq r - 1$ *and* $a + b + c \leq 2r - 2$.

Remark 8 *A slight bit of algebra gives a more elegant formula for θ. First note the following Lemma.*

Lemma 13 $\Delta_{m+p}\Delta_{n+p-1} - \Delta_m\Delta_{n-1} = \Delta_{m+n+p}\Delta_{p-1}$

Proof: We take $\Delta_k = (x^{k+1} - x^{-k-1})/x - x^{-1})$. Then

$$\Delta_{m+p}\Delta_{n+p-1} - \Delta_m\Delta_{n-1} = (x - x^{-1})^{-2}$$

$$[(x^{m+p+1} - x^{-m-p-1})(x^{n+p+1} - x^{-n-p+1}) - (x^{m+1} - x^{-m-1})(x^n - x^{-n})]$$

$$= (x - x^{-1})^{-2}[x^{m+n+2p+1} - x^{m-n+1} - x^{n-m-1} + x^{-m-n-2p-1}$$

$$- x^{m+n+1} + x^{m-n+1} + x^{-m+n-1} - x^{-m-n-1}]$$

$$= (x - x^{-1})^{-2}[x^{m+n+2p+1} - x^{m+n+1} - x^{-m-n-1} + x^{-m-n-2p-1}]$$

$$= (x - x^{-1})^{-2}[x^{m+n+p+1} - x^{-m-n-p-1}][x^p - x^{-p}] = \Delta_{m+n+p}\Delta_{p-1}$$

∎

Lemma 14 $\rho(m, n, p) = \dfrac{\Delta_{m+n+p}\Delta_{p-1}}{\Delta_{m+p-1}\Delta_{n+p-1}}.$

Proof:

$$
\begin{aligned}
\rho(m, n, p) &= d - \mu_{m+p-1} - \frac{\Delta_m\Delta_{n-1}}{\Delta_{m+p-1}\Delta_{n+p-1}} \\
&= \frac{d\Delta_{m+p-1}\Delta_{n+p-1} - \mu_{m+p-1}\Delta_{m+p-1}\Delta_{n+p-1} - \Delta_m\Delta_{n-1}}{\Delta_{m+p-1}\Delta_{n+p-1}}.
\end{aligned}
$$

Note that $\mu_{k+1} = \Delta_k/\Delta_{k+1}$. Hence

$$
\begin{aligned}
\rho(m, n, p) &= \frac{d\Delta_{m+p-1}\Delta_{n+p-1} - \Delta_{m+p-2}\Delta_{n+p-1} - \Delta_m\Delta_{n-1}}{(\Delta_{m+p-1})(\Delta_{n+p-1})} \\
&= \frac{(d\Delta_{m+p-1} - \Delta_{m+p-2})\Delta_{n+p-1} - \Delta_m\Delta_{n-1}}{(\Delta_{m+p-1})(\Delta_{n+p-1})} \\
&= \frac{\Delta_{m+p}\Delta_{n+p-1} - \Delta_m\Delta_{n-1}}{(\Delta_{m+p-1})(\Delta_{n+p-1})} \\
\therefore \quad \rho(m, n, p) &= \frac{\Delta_{m+n+p}\Delta_{p-1}}{\Delta_{m+p-1}\Delta_{n+p-1}}
\end{aligned}
$$

∎

Corollary 2 *Let* $\Delta_n!$ *denote the product* $\Delta_n! = \Delta_n\Delta_{n-1}\Delta_{n-2}\ldots\Delta_1$. *Then*

$$
\theta(a,b,c) \;=\; \left[\;\begin{matrix}\overset{a}{\underset{c}{\bullet\!\!\!-\!\!\!b\!\!\!-\!\!\!\bullet}}\end{matrix}\;\right] \;=\; \left[\;\begin{matrix}p\\ \underset{m\quad n}{\bigcirc\!\!\!-\!\!\!\bigcirc\!\!\!-\!\!\!\bigcirc}\end{matrix}\;\right]
$$

$$
= \frac{\Delta_{m+n+p}!\,\Delta_{n-1}!\,\Delta_{m-1}!\,\Delta_{p-1}!}{\Delta_{n+p-1}!\,\Delta_{m+p-1}!\,\Delta_{m+n-1}!} = \frac{(-1)^{m+n+p}[m+n+p+1]!\,[n]!\,[m]!\,[p]!}{[n+p]!\,[m+p]!\,[m+n]!}.
$$

Proof: By 1, we have that

$$
\theta(a,b,c) = \left(\prod_{j=1}^{p}\rho(m,n,j)\right)\Delta_{m+n}.
$$

Hence, by 3.15,

$$
\begin{aligned}
\theta(a,b,c) &= \prod_{j=1}^{p}\left[\frac{\Delta_{m+n+j}\Delta_{j-1}}{\Delta_{m+j-1}\Delta_{n+j-1}}\right]\Delta_{m+n}\\[2mm]
&= \frac{(\Delta_{m+n+p}\Delta_{m+n+p-1}\cdots\Delta_{m+n})\Delta_{p-1}!}{(\Delta_{m+p-1}\Delta_{m+p-2}\cdots\Delta_m)(\Delta_{n+p-1}\Delta_{n+p-2}\cdots\Delta)}\\[2mm]
&= \frac{\Delta_{m+n+p}!\,\Delta_{n-1}!\,\Delta_{m-1}!\,\Delta_{p-1}!}{\Delta_{m+p-1}!\,\Delta_{n+p-1}!\,\Delta_{m+n-1}!}.
\end{aligned}
$$

In the above formula we can take $\Delta_{-1} = \Delta_0 = 1$. With these definitions it applies for all non-negative triples m, n, p (the boundary case Δ_{-1} was treated in the basis of the recursion.) The fact that $[n] = (-1)^{n-1}\Delta_{n-1}$ holds for $n \geq 1$ and an easy exercise in parity rewrites the formula in terms of the quantum factorials as displayed: indeed, we get

$$
\Delta_{m+n+p}! = (-1)^{(m+n+p)(m+n+p+1)/2}[m+n+p+1]!
$$

$$
\Delta_{m+n-1}! = (-1)^{(m+n-1)(m+n)/2}[m+n]!
$$

$$
\Delta_{m-1}! = (-1)^{(m-1)m/2}[m]!
$$

It follows that the exponent of -1 after collecting terms is

$$
2m^2 + 2n^2 + 2p^2 + 2mn + 2np + 2pm - m - n - p \equiv m + n + p \ (\mathrm{mod}\ 2).
$$

Therefore, we get

$$\frac{\Delta_{m+n+p}!\Delta_{n-1}!\Delta_{m-1}!\Delta_{p-1}!}{\Delta_{n+p-1}!\Delta_{m+p-1}!\Delta_{m+n-1}!} = \frac{(-1)^{m+n+p}[m+n+p+1]![n]![m]![p]!}{[n+p]![m+p]![m+n]!}.$$

∎

This evaluation is implicit in [Kau91],[Kau90a] and has also been noted in the Temperley-Lieb context by [Lic92] and [MV92]. In *Chapter 8* we will discuss how these and similar formulas arise as the q-analogs to simple counting processes in the classical case ($q = \pm 1$).

In evaluating the networks that appear in the proof of the recoupling formula of the next section we are led to networks having vertices of the type

where $a^*, b^*, c^* \leq r - 1$ and $a^* + b^* + c^* \geq 2r - 2$. We must prove

Lemma 15 *The evaluation of such networks (described above) is zero.*

Proof: Let $m^* = (a^* + b^* - c^*)/2$, $n^* = (b^* + c^* - a^*)/2$, $p^* = (a^* + c^* - b^*)/2$ and let $m \leq m^*$, $n \leq n^*$, $p \leq p^*$ be any choices of non-negative integers satisfying $m + n + p = r - 1$. We get (network evaluations at the r-th root of unity)

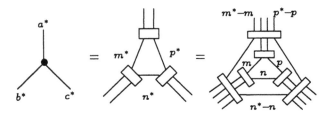

Expanding every projector but the redundant three we have introduced, every term which has outside connections distinct from the terms involving

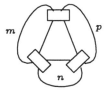

has turnbacks and thus is null. Therefore, for appropriate $= \lambda_i' s$ we have

$$
\begin{array}{c} a^* \\ \bigwedge \\ b^* \quad c^* \end{array} = \left(\sum_i \lambda_i \right) \quad
$$

,

and this is null since the last network is $Net(m, n, p)$ with $m + n + p = r - 1$. This proves the Lemma. ∎

Chapter 7

Recoupling Theory Via Temperley-Lieb Algebra

7.1 Recoupling Theorem

We now state and prove the basic recoupling theorem for $q = e^{i\pi/r}$. To this purpose let ADM_q denote *the set q-admissible triples* $\{a, b, c\}$. A triple of non-negative integers $\{a, b, c\}$ is said to be *q-admissible* if

(i) $a + b + c \equiv 0 \qquad (mod\ 2)$

(ii) $a + b - c, \quad b + c - a, \quad c + a - b$ are each ≥ 0.

(iii) $a + b + c \leq 2r - 4$.

The first two conditions of q-admissibility just correspond to the general admissibility for a 3-vertex. Condition (iii) corresponds to the criterion of *Proposition 8*.

Theorem 2 (Recoupling Theorem) *Let a, b, c, d, j be non-negative fixed integers such that $\{a, b, j\}$ and $\{c, d, j\}$ are q-admissible triples. Then there* *exist unique real numbers* $\alpha_i, 0 \leq i \leq r-2$, *such that*

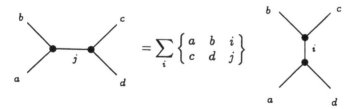

$$ \text{[diagram]} = \sum_{\substack{i \\ \{b,c,i\} \in ADM_q \\ \{a,d,i\} \in ADM_q}} \alpha_i \quad \text{[diagram]}. $$

All network evaluations are at $q = e^{\pi i/r}$.

Remark 9 *The equality means that both sides are interchangeable in all bracket calculations involving these network sums inside of large networks. We have written the formula to emphasize q-admissibility. It will be convenient to note the dependence of* α_i *on* i, j, a, b, c, d *by writing*

$$ \alpha_i = \left\{ \begin{matrix} a & b & i \\ c & d & j \end{matrix} \right\}_q . $$

Sometimes the subscript q is dropped. The symbol $\left\{ \begin{matrix} a & b & i \\ c & d & j \end{matrix} \right\}_q$ *is a quantum q-6j symbol [KR88]. We will not explain here precisely how our q-6j symbols, defined via the Temperley-Lieb algebra, are related to the q-6j symbols coming from the* $SL(2)_q$ *quantum group [KR88]. (But compare [Piu92].)*

Thus we shall write the Recoupling Theorem in the form

$$ \text{[diagram]} = \sum_i \left\{ \begin{matrix} a & b & i \\ c & d & j \end{matrix} \right\} \quad \text{[diagram]} $$

where it is assumed that all of the triples satisfy the admissibility conditions:

$$ \{a,b,j\}, \{c,d,j\}, \{b,c,i\} \quad \text{and} \quad \{a,d,i\} \in ADMq. $$

In order to prove this theorem it is convenient to consider first the collection of all tangles of the form below

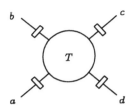

Let $T\begin{bmatrix} a & b \\ c & d \end{bmatrix}$ denote the collection of such tangles. The input lines are parallel cables of size a, b, c, d respectively and each cable runs through a projector as shown in the diagram above. Each tangle is a functional on tangles $\hat{T}$ "dual" to it in the plane (i.e. drawn in the exterior space of the tangle) and we have an inner product

$$< \, , \, >: T\begin{bmatrix} a & b \\ c & d \end{bmatrix} \times \hat{T}\begin{bmatrix} a & b \\ c & d \end{bmatrix} \longrightarrow \mathbb{C}$$

determined by bracket evaluation of the connection of two such tangles. Call two elements of $T\begin{bmatrix} a & b \\ c & d \end{bmatrix}$ *equal* if they are equal as functionals on the dual tangles $\hat{T}\begin{bmatrix} a & b \\ c & d \end{bmatrix}$. This generalizes the equality in the statement of the recoupling theorem.

Note that $T\begin{bmatrix} a & b \\ c & d \end{bmatrix}$ is a vector space over $\mathbb{C}$ (via addition of tangles) and that in any tangle the replacement $\left(\asymp \ = \ A \smile\hspace{-0.3em}\frown \ + \ A^{-1} \right)($ is valid since this does not effect the bracket evaluation for $A = \sqrt{q}, q = c^{i\pi/r}$.

Note also that since

$$\text{(diagram)} \quad = \quad \text{(diagram)} \quad ,$$

it follows that the recoupling theorem is a statement of a linear identity in the vector space $T\begin{bmatrix} a & b \\ c & d \end{bmatrix}$.

In fact, we shall prove that the set of tangles of the form

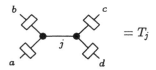

with $\{a, b, j\}, \{c, d, j\} \in ADMq$ is a *basis* for the space $T\begin{bmatrix} a & b \\ c & d \end{bmatrix}$. The same argument will show that the tangles

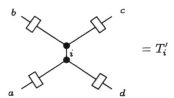

with $\{b, c, i\}, \{a, d, i\} \in ADMq$ are also a basis for $T\begin{bmatrix} a & b \\ c & d \end{bmatrix}$. The Recoupling Theorem then follows via change of basis.

Lemma 16 *Let* $B = \{T_j | \{a, b, j\}, \{c, d, j\} \in ADMq\}$ *denote the set of elements of* $T\begin{bmatrix} a & b \\ c & d \end{bmatrix}$ *as described above. Then B is linearly independent over* $\mathbb{C}$.

Proof: Note that

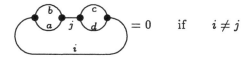

by Lemma 7. Apply this form of closure to each term of any linear dependence relation $\sum_j c_j T_j = 0$

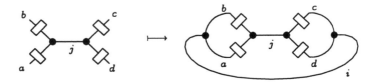

and conclude that for

$$\{a, b, j\} \quad \text{and} \quad \{c, d, j\} \in ADMq$$

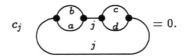

However, 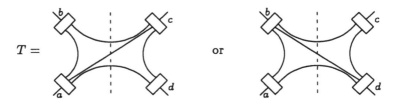 $= \dfrac{\theta(a, b, j)\theta(c, d, j)}{\Delta_j}$ and the admissibility cri-

teria on (a, b, j) and (c, d, j) imply (by Proposition 8) that $\theta(a, b, j) \neq 0$, $\theta(c, d, j) \neq 0$. *Also*, $\Delta_j \neq 0$, completing the proof of linear independence. ∎

Lemma 17 *Let B be as in the previous Lemma. Then B is a basis for the space $T\begin{bmatrix} a & b \\ c & d \end{bmatrix}$.*

Proof: By the previous Lemma it suffices to show that B spans $T\begin{bmatrix} a & b \\ c & d \end{bmatrix}$. By the remarks prior to that Lemma it suffices to show that any flat tangle $T \in T\begin{bmatrix} a & b \\ c & d \end{bmatrix}$ can be expressed in terms of B. Such a flat tangle has the form shown below.

$$T =$$ or

with the multiplicities of internal lines adding up to a, b, c, d respectively. The dotted line in the diagrams above is intersected transversely by n arcs for some n, forming a parallel cable that we denote (rotating by $90°$)

by $- - -|\overset{n}{\underset{}{-}} - - = 1n$ Regard this cable as the identity element in the

Temperley Lieb algebra T_n (with $q = e^{i\pi/r}$).

Then $1n = \overset{|n}{\boxed{}} + R_n$ where $\overset{|n}{\boxed{}}$ is a projector and R_n is a linear

combination of cables with turn-arounds in them (a product of $U_i's$ for each

summand). Furthermore, since $\boxed{\overset{r-1}{}} = 0$, we can reduce the problem of

rewriting to the case $0 \leq n \leq r - 2$ by successive rewriting of bundles of $r - 1$ lines: $1_{r-1} = R_{r-1}$. For $n \leq r - 1$ we have

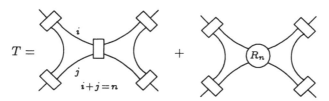

$$T = \quad + \quad$$

with the vertical cross-sections of each term in R_n less than n. This writes T as a linear combination of tangles of the form with $0 \leq j \leq r - 2$. Since $= 0$ as a functional on

tangles unless $(a, b, j) \in ADMq$. This completes the proof that B spans $\mathcal{T}\begin{bmatrix} a & b \\ c & d \end{bmatrix}$. ∎

7.2   The case of General q

Remark 10 *The same method as employed in the proof of above gives the recoupling expansion for generic q (q not a root of unity). In the generic case the summation is over all admissible triples (as opposed to q-admissible) where a triple $\{a, b, c\}$ is said to be admissible if*

  (i) $a + b + c \equiv 0 \pmod 2$
  (ii) $a + b - c, b + c - a$ and $c + a - b$ are each ≥ 0.

An example will suffice to illustrate the point: In the generic case, we expact to find α, β, γ such that

Explicitly, this equation becomes

Here we have eliminated the exterior leads on the four outer symmetrizers in order to emphasize that all terms will be seen as tangles flanked by these four symmetrizers. Due to the fact that turn-backs into any one of the four will produce a null term, we see that the only available tangles are the following ones:

Upon expanding the middle symmetrizer we have

$$\text{\includegraphics} = \text{\includegraphics} - \frac{1}{d} \text{\includegraphics}$$

and

$$\text{\includegraphics} = r \text{\includegraphics} + s \text{\includegraphics} + \text{\includegraphics} .$$

where the specific values for r and s can be obtained from the expansion of

$$\boxed{\begin{array}{c} 4 \\ \end{array}} = f_3. \ (\textit{One finds that } r = \mu_1\mu_2^2\mu_3(2 - \mu_1 d) \textit{ and } s = -\mu_2(1 - \mu_2\mu_3).)$$

Collecting terms, we find that equation (*) becomes

$$\text{\includegraphics} = \left(\alpha - \frac{\beta}{d} + s\gamma\right) \text{\includegraphics} + (\beta + r\gamma) \text{\includegraphics} + \gamma \text{\includegraphics} .$$

Thus

$$\begin{aligned} \gamma &= 1 \\ \beta + r\gamma &= 0 \\ \alpha - \beta/d + s\gamma &= 0 \end{aligned}$$

is the system of equations to be solved. Since this is triangular, with ones on the diagonal, it always has a solution.

The difference between the generic case and the case at α root of unity is also well-illustrated by this example. For example, suppose $q = e^{i\pi/5}$ so that 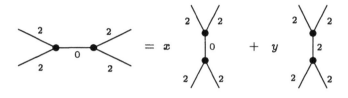 $= f_3 = 0$. Then q-admissibility entails $a + b + c \leq 2 \cdot 5 - 4 = 6$, whence the recoupling formula becomes

or

It is the vanishing of the symmetrizer that resolves this equation via

Thus

$$\Rightarrow \quad -s = x - \tfrac{y}{d}, \quad -r = y$$

$$\Rightarrow \quad x = -\tfrac{r}{d} - s$$

$$y = -r$$

giving the recoupling coefficients at this root of unity. The proof of Recoupling Theorem encodes this rewriting technique for the case of any $e^{i\pi/r} = q$.

7.3   Orthogonality and Pentagon Identities

As we have remarked before, it is convenient to regard the input values a, b, c, d, j in the recoupling, not as fixed, but as parameters of a six parametric function

$$\{\ \}_q : \{0, 1, 2, \ldots, r-2\}^6 \longrightarrow \mathbb{R}$$

given by $\left\{ \begin{matrix} a & b & i \\ c & d & j \end{matrix} \right\}_q = \alpha_i$ where α_i is the coefficient of

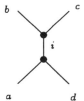

in the recoupling expansion. The values $\left\{ \begin{matrix} a & b & i \\ c & d & j \end{matrix} \right\}_q$ are called q-deformed $6j$ coefficients, or $q - 6j$ coefficients for short. Two of their properties hold the key to the 3-manifold invariant discussed in Chapter 10. The orthogonality and pentagon (Biedenharn-Elliot) identities, which we treat next. The Recoupling Theorem also provides the key for an explicit (network evaluation) formula for the $q - 6j$ coefficients.

In the following, we shall state and prove the orthogonality and pentagon identities for the case $q = e^{i\pi/r}$ and q-admissibility. The corresponding generic theorems are also true, and easily proved by the same methods.

Proposition 9 *[Orthogonality Identity]:*

$$\sum_{i=0}^{r-2} \left\{ \begin{matrix} a & b & i \\ c & d & j \end{matrix} \right\}_q \left\{ \begin{matrix} d & a & k \\ b & c & i \end{matrix} \right\}_q = \left\{ \begin{matrix} 0 & if & k \neq j \\ 1 & if & k = j \end{matrix} \right.$$

Proof: This identity follows from a double use of the Recoupling Theorem:

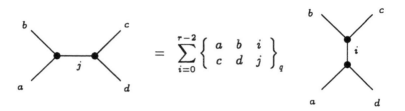

$$\text{(network)} = \sum_{i=0}^{r-2} \left\{ \begin{matrix} a & b & i \\ c & d & j \end{matrix} \right\}_q \text{(network)} \; .$$

Rotate the last network clockwise, and apply the Recoupling Theorem again

$$\text{(network)} = \sum_{k'=0}^{r-2} \left\{ \begin{matrix} d & a & k' \\ b & c & i \end{matrix} \right\}_q \text{(network)} \; .$$

Now rotate back, and use this identity in the first one.

$$\text{(network)} \quad =$$

$$= \sum_{i=0}^{r-2} \left\{ \begin{matrix} a & b & i \\ c & d & j \end{matrix} \right\}_q \; \sum_{k'=0}^{r-2} \left\{ \begin{matrix} d & a & k' \\ b & c & i \end{matrix} \right\}_q \text{(network)}$$

Specializing, we find, for fixed k,

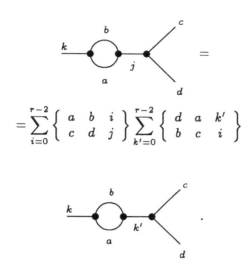

$$= \sum_{i=0}^{r-2} \left\{ \begin{array}{ccc} a & b & i \\ c & d & j \end{array} \right\} \sum_{k'=0}^{r-2} \left\{ \begin{array}{ccc} d & a & k' \\ b & c & i \end{array} \right\}$$

The result then follows directly from Lemma 7 and the non-vanishing results of Lemma 4 and Proposition 8. ∎

Proposition 10 *[Pentagon Identity (Biedenharn-Elliot)]:*

$$\sum_{m=0}^{r-2} \left\{ \begin{array}{ccc} a & i & m \\ d & e & j \end{array} \right\}_q \left\{ \begin{array}{ccc} b & c & l \\ d & m & i \end{array} \right\}_q \left\{ \begin{array}{ccc} b & l & k \\ e & a & m \end{array} \right\}_q =$$
$$= \left\{ \begin{array}{ccc} b & c & k \\ j & a & i \end{array} \right\}_q \left\{ \begin{array}{ccc} k & c & l \\ d & c & j \end{array} \right\}_q .$$

Proof: Consider the pentagonal diagram:

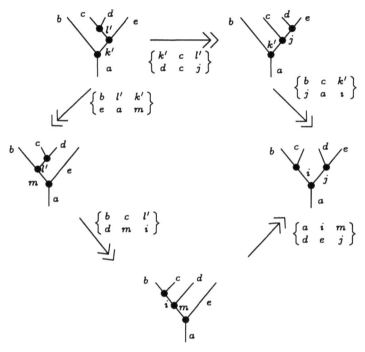

There are two ways to express the network on the right in terms of the network at the upper left. One way involves two applications of the Recoupling Theorem, the other requires three applications. (To save repetition, we write $\{\ \}$ for $\{\ \}_q$.)

(ii)

$$\begin{array}{l} = \sum_{k'=0}^{r-2} \left\{ \begin{array}{ccc} b & c & k' \\ j & a & i \end{array} \right\} \end{array}$$

$$= \sum_{k'=0}^{r-2} \sum_{l'=0}^{r-2} \left\{ \begin{array}{ccc} b & c & k' \\ j & a & i \end{array} \right\} \left\{ \begin{array}{ccc} k' & c & l' \\ d & e & j \end{array} \right\}$$

.

(iii)

$$\text{(diagram)} = \sum_{m=0}^{r-2} \left\{ \begin{array}{ccc} a & i & m \\ d & e & j \end{array} \right\}$$

$$\sum_{l'=0}^{r-2} \left\{ \begin{array}{ccc} b & c & l' \\ d & m & i \end{array} \right\} \sum_{k'=0}^{r-2} \left\{ \begin{array}{ccc} b & l' & k' \\ e & a & m \end{array} \right\} \quad \text{(diagram)}$$

Therefore for a particular choice of contraction we obtain

$$\sum_{k'=0}^{r-2} \sum_{l'=0}^{r-2} \left\{ \begin{array}{ccc} b & c & k' \\ j & a & i \end{array} \right\} \left\{ \begin{array}{ccc} k' & c & l' \\ d & e & j \end{array} \right\} \quad \text{(diagram)}$$

$$= \sum_{m=0}^{r-2} \sum_{l'=0}^{r-2} \sum_{k'=0}^{r-2} \left\{ \begin{array}{ccc} a & i & m \\ d & c & l \end{array} \right\} \left\{ \begin{array}{ccc} b & c & l \\ d & m & i \end{array} \right\}$$

$$\left\{ \begin{array}{ccc} b & l' & k' \\ e & a & m \end{array} \right\} \quad \text{(diagram)} \quad .$$

By using Lemma 7 twice, we get

$$\left(\sum_{m=0}^{r-2} \left\{ \begin{array}{ccc} a & i & m \\ d & e & j \end{array} \right\} \left\{ \begin{array}{ccc} b & c & l \\ d & m & i \end{array} \right\} \left\{ \begin{array}{ccc} b & l & k \\ e & a & m \end{array} \right\} - \right.$$

$$\left. - \left\{ \begin{array}{ccc} b & c & k \\ j & a & i \end{array} \right\} \left\{ \begin{array}{ccc} k & c & l \\ d & e & j \end{array} \right\} \right)$$

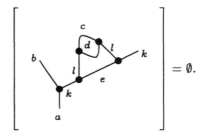

$$= \emptyset.$$

The result follows because there are non-null, choices for the network. Indeed,

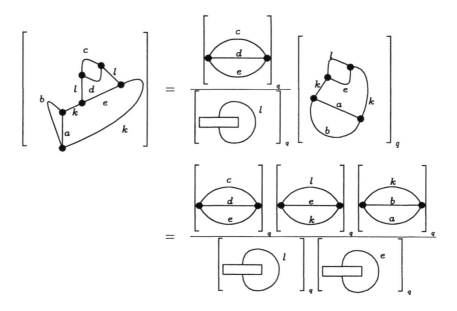

There are choices for a, b, c, d, e, k, l making the above evaluation non-null. This completes the proof. ∎

We finish by obtaining a formula for the $q - 6j$:

Proposition 11

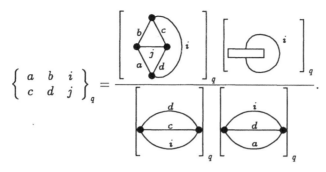

$$\left\{ \begin{array}{ccc} a & b & i \\ c & d & j \end{array} \right\}_q = \frac{\left[\vphantom{\Big|} \right] \left[\vphantom{\Big|} \right]}{\left[\vphantom{\Big|} \right] \left[\vphantom{\Big|} \right]}.$$

This formula holds under conditions of general admissibility for generic q, and under conditions of q-admissibility (see the beginning of section 7) for q of the form $e^{i\pi/r}$.

Proof: The formula is an immediate consequence of the Recoupling Theorem and Lemma 7:

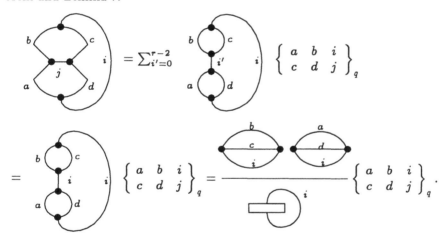

This completes the proof. ∎

Chapter 8

Chromatic Evaluations and the Tetrahedron

8.1 Exact Formulas in a Special Case

The purpose of this Chapter is to show how to derive the exact formulas for the theta and tetrahedral nets in the special case $A = -1$ by a simple counting argument. These formulas are expressed (for $A = -1$) in terms of ordinary factorials. The general formulas that we need for arbitrary A are obtained by replacing integers in these formulas by "quantum integers",

$$[n] = \frac{A^{2n} - A^{-2n}}{A^2 - A^{-2}} = (-1)^{n-1}\Delta_{n-1}.$$

Note that $[n] = A^{2(n-1)}(1 + A^{-4} + \cdots + (A^{-4})^{n-1}) = A^{2(n-1)}\{n\}$ so that $[n](A = \pm 1) = n$. For our objectives here it is more convenient to use this definition of quantum integer. The result is a shift by the factor $A^{2(n-1)}$ from the use of $\{n\}$. Thus we now write

$$\boxed{}^{\,n} = \frac{A^{n(n-1)}}{[n]!} \sum_{\sigma \in S_n} (A^{-3})^{t(\sigma)} \quad \boxed{\hat{\sigma}}^{\,n}$$

since $[n]! = A^{\sum_k^n - 1\, 2(k-1)} \prod_{k=1}^{n} \left(\frac{1 - A^{-4k}}{1 - A^{-4}}\right) = A^{n(n-1)}\{n\}!$ where $\{n\}!$ denotes the version of quantum factorial used in Chapter 1.

In the special case $A = -1$ the projector becomes

$$\overset{|n}{\boxed{}} = \frac{1}{n!} \sum_{\sigma \in S_n} (-1)^{t(\sigma)} \overset{|n}{\boxed{\hat{\sigma}}}$$

where $\asymp = -\overset{\cup}{\underset{\cap}{}} - \Big)$ $\Big($ and the loop value is $d = -(-1)^2 - (-1)^{-2} = -2$.

Here $\asymp = \asymp$ so that we can write $\times = -\overset{\cup}{\underset{\cap}{}} - \Big)$ $\Big($ with no crossing

indication and the projector sum is a sum over all permutations of crossing diagrams that represent these permutations. Thus

$$\overset{||}{\boxed{}} = \frac{1}{2!} \left(\;||\; - \times \right) = \frac{1}{2} \left(\;||\; + \overset{\cup}{\underset{\cap}{}} + \right) (\;) = \;||\; + \frac{1}{2} \overset{\cup}{\underset{\cap}{}}$$

for $d = -2$.

While the knot theory is of no interest for $A = -1$, the recoupling theory is non-trivial. In fact at this value, the recoupling theory is equivalent to the recoupling theory of classical quantum angular momentum for the group $SU(2)$. We shall not go into the details of the relationship and the translations between these points of view and the theory of quantum groups. For this the reader is referred to [Kau90c], [Kau91], [Kau92] and [Piu92]. These same references explain how the recoupling theory at $A = -1$ is equivalent to the Penrose spin network theory. The method of evaluation discussed here is a generalization of the chromatic method of spin network evaluation [Mou79].

8.2 Tensorial Formalism

Let V be a vector space over $\mathbb{C}$ of dimension N, with basis $B = \{e_1, e_2, \ldots, e_N\}$. For all $N \geq n$ we can define a projector

$$P_n^N = \frac{1}{n!} \sum_{\sigma \in Sn} (-1)^{t(\sigma)} \hat{\sigma}$$

where

$$P_n^N : V \otimes V \otimes \cdots \otimes V \longrightarrow V \otimes V \otimes \ldots \otimes V$$

(n tensor factors) via

$$\hat{\sigma} : V^{\otimes n} \longrightarrow V^{\otimes n}$$
$$\hat{\sigma}(e_{i_1} \otimes \cdots \otimes e_{i_n}) = e_{i_{\sigma(1)}} \otimes \cdots \otimes e_{i_{\sigma(n)}}$$

In the matrix form we can write

$$\hat{\sigma} = \delta^{a_1}_{b_{\sigma(1)}} \delta^{a_2}_{b_{\sigma(2)}} \cdots \delta^{a_n}_{b_{\sigma(n)}} = [\hat{\sigma}]^{a_1 \cdots a_n}_{b_1 \cdots b_n}$$

where $\delta^i_j = \left\{ \begin{array}{ll} 1 & \text{if} \quad i = j \\ 0 & \text{if} \quad i \neq j \end{array} \right.$ and $1 \leq i, j \leq N$.

By letting δ^i_j correspond to an arc or segment as shown below

$$\delta^i_j \quad \longleftrightarrow \quad \Big|^i_j$$

we can retain our previous diagrammatic technique. For example,

$$P_2^N = \frac{1}{2!} \left(\Big|\Big| - \times \right)$$

and

$$(P_2^N)^{ab}_{cd} = \frac{1}{2!}(\delta^a_c \delta^b_d - \delta^a_d \delta^b_c).$$

We take the trace of a matrix as the value of its strand closure. Thus

$$tr \left(\begin{array}{c} \Big|^i \\ \boxed{A} \\ \Big|_j \end{array} \right) = \boxed{A} \!\!\supset = \sum_i A^i_i.$$

In general a network can be decomposed as a contraction of many Kronecker deltas:

$$\boxed{A} \!\!\supset \quad \longleftrightarrow \quad \sum_{i,j} \quad \overset{i \bullet}{\boxed{A}} k\bullet \\ {}_{j \bullet} \quad = \sum_{i,j,k} A^i_j \delta^i_k \delta^k_j = \sum_i A^i_i = tr(A).$$

It is easy to see that the value is independent of the particular choice of decomposition.

With these conventions, we have $tr(\delta_j^i) = N$, the value of a single loop. Similarly, $tr(P_2^N) = \frac{1}{2!}\left(\text{⬭} - \text{⬭} \right) = \frac{1}{2}(N^2 - N)$.

In general, in evaluating any loop (possibility with self-crossings) at positive dimension N, the value is simply equal to N. For the bracket evaluation at $A = -1$ the same fact is true! For we have $< \text{⤬} >= < \text{⤬} >$ and

$$< \text{⤬} >= -A^3 < \text{⌣} >= -(-1)^3 < \text{⌣} >=< \text{⌣} > .$$

Thus the immersed curves for $A = -1$ can be subjected to arbitrary flat Reidemeister moves and undone. For example:

$$\left\langle \text{⬭} \right\rangle = \left\langle \text{⬭} \right\rangle = \left\langle \bigcirc \right\rangle = -2.$$

$$\left\langle \text{8} \right\rangle = -\left\langle \text{8} \right\rangle - \left\langle \bigcirc \right\rangle = -4 + 2 = -2.$$

Terminology. We shall refer to network evaluations at dimension N. *It is understood that dimension $N = -2$ refers to the bracket evaluation for $A = -1$.*

The proceeding discussion proves the following.

Theorem 3 *Let $\mathcal{N}$ be a (spin) network immersed in the plane (i.e. a closed network composed of projectors and possibly containing crossovers among its lines). Let $||\mathcal{N}||(N)$ denote the value of this network computed as function of N for positive N. Then $< \mathcal{N} > (-1) = ||\mathcal{N}||(-2)$ where $< \mathcal{N} > (-1)$ denotes the bracket evaluation of this network at $A = -1$.*

Proof: Expand all the projectors as sums of permutations. Note that $< \mathcal{N} > (-1)$ assigns each loop the value (-2) and so does $||\mathcal{N}||(-2)$. Hence the computations are identical in this form. ∎

The value of this Theorem lies in the fact that there are other ways to compute $||\mathcal{N}||$ in positive dimensions.

In particular, for positive N the matrix entries of P_n^N are zero unless the input indices are all distinct and the output indices are a permutation of the input indices:

Lemma 18 $(P_n^N)_{b_1 b_2 \cdots b_n}^{a_1, a_2 \cdots a_n} = 0$ *unless the values of* $a_1, \ldots, a_n$ *are all distinct and the values of* $b_1, b_2, \ldots, b_n$ *are a permutation of* $a_1, \ldots, a_n$. *If* $a_1, \ldots, a_n$ *are distinct and* $b_{\sigma(i)} = a_i$ *for* $i = 1, \ldots, n$ *and some* $\sigma \in S_n$, *then*

$$(P_n^N)_{b_1 \cdots b_n}^{a_1, \cdots a_n} = (-1)^{t(\sigma)}/n!$$

Therefore $tr(P_n^N) = \frac{N(N-1)\cdots(N-n+1)}{n!} = \binom{N}{n}$ *where* $\binom{N}{n} = \frac{N!}{n!(N-n)!}$
is a binomial coefficient.

Proof: This follows directly from the definition of P_n^N as an alternating sum of permutation morphisms on $V^{\otimes n}$. ∎

Remark 11 *From the Lemma, we have*

$$\left\| \, n \, \rightbox \, \right\| = \frac{N(N-1)\cdots(N-n+1)}{n!} = \binom{N}{n}$$

whence for $N = -2$,

$$\binom{-2}{n} = \frac{-(2)(-3)\cdots(-2-n+1)}{n!}$$

$$= (-1)^n (n+1)!/n! = (-1)^n (n+1).$$

On the other hand,

$$\left\langle \, n \, \rightbox \, \right\rangle (A) \;\; = \Delta_n(-A^2) = \frac{(-A^2)^{n+1} - (-A^2)^{-n-1}}{(-A^2) - (-A^2)^{-1}}$$

$$= (-1)^n \frac{A^{2n+2} - A^{-2n-2}}{A^2 - A^2} = (-1)^n [n+1]$$

Thus $\left\langle \, n \, \rightbox \, \right\rangle (A = -1) = (-1)^n (n+1).$

8.3 A Heuristic Correspondence on the θ-Net

The previous example suggests the correspondence

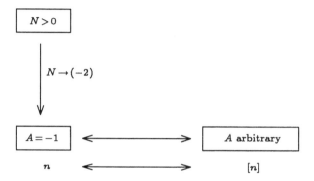

By computing the special case $A = -1$ by using postive dimension calculations, *we may transfer the result to the general case of arbitrary A by replacing integers n by quantum integers [n]*. While we shall not formalise this procedure here, we shall illustrate its use for both theta and tetrahedral evaluations.

First consider the theta evaluation. This is equivalent to computing a net of the form

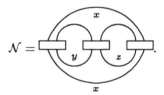

Consider $||\mathcal{N}||(N)$ with $N \geq n$. Each projector must, in a given non-vanishing term of the contraction, receive and transmit (permuted) a distinct set of "colors" from the set $\{1, 2, \ldots, N\}$. Thus we must count all admissible colorings of the net. Since a color that enters the bar of a projector must also leave it, it follows that there are loops of constant color in the net and that *no loop can go through a given bar more that once*. A configuration of such loops is specifically determined by a choice of permutation at each bar. For example $(x = y = z = 1)$

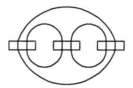

is a valid configuration of loops and it contributes $N(N-1)(N-2)$ distinct colorings, since each pair of these loops shares a bar. The contribution of this configuration to the network evaluation is therefore $N(N-1)(N-2)/(2!)^3$ since each bar contributes a factor of $(2!)$.

In this case there are no other valid configurations. For example

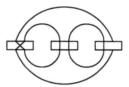

has a loop re-entering the left-hand bar – forcing a repetition of color at this bar.

On the other hand, consider $x = 2, y = z = 1$:

(a) 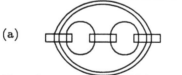   (b)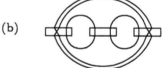

Now there are two valid loop configurations, but only the three basic loops. The outer form of loop carries a cable of two loops and these can assume the configurations a) and b) depending upon internal permutation.

Thus we see by direct count that the value of the net $\mathcal{N}$ at positive N is given by the formula

$$||\mathcal{N}||(N) = \frac{x!y!z!(N(N-1)\cdots(N-x-y-z+1))}{(x+y)!(x+z)!(y+z)!}.$$

The factorials in the denominator correspond to the sizes $(x+y), (x+z), (y+z)$ of the projectors. The factorials of x, y and z in the numerator count interior cable multiplicities for the three loop forms of size x, y, z. The factor

$$N(N-1)\cdots(N-x-y-z+1)$$

is the number of ways to color $x + y + z$ pairwise distinct loops.

Note that each individual colored loop contribution is multiplied by the product of signs of the permutations at bars. since this sign is the parity of the number of loop intersections, it is always $+1$ since each loop is non self-intersecting in the plane. Thus our formula represents the sum of these contributions.

Now taking $A = -1, N = -2$ we have

$$<\mathcal{N}>(A = -1) = \frac{x!y!z!(-1)^{x+y+z}(x+y+z+1)!}{(x+y)!(x+z)!(y+z)!}$$

This completes a rigorous derivation of the network value for $A = -1$. If we replace $x!$ by $[x]!$ with $[x] = (-1)^{x-1}\Delta_{x-1}$, then it becomes

$$<\mathcal{N}>(A) = \frac{[x]![y]![z]!(-1)^{x+y+z}[x+y+z+1]!}{[x+y]![x+z]![y+z]!}$$

This is exactly the formula we derived in Chapter 6 for the general theta net.

8.4   Chromatic Evaluation: General Case

For our purposes, the most useful generalization of this computation will be to a formula for the positive N evaluation of a planar closed network (i.e. embedded in the plane) with trivalent nodes. Therefore let G be a trivalent plane graph with integer labels on its edges such that each node is *admissible* (If a, b, c are labels incident to the edge then $a + b - c, a + c - b, b + c - a$ are all positive and even.). Note that we can draw a network of projectors equivalent to this graph by the following algorithm.

1) Draw a Jordan curve within each region delineating its boundary.

2) Remove the original graph, placing a projector across the region boundaries parallel to each edge.

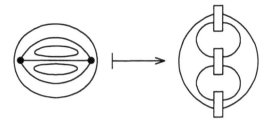

3) Label the resulting edges of the projector network so that the sum of labels going into a side of a projector is equal to the label on the original graphs.

$\mathcal{N}(G)$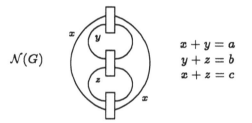

$$x + y = a$$
$$y + z = b$$
$$x + z = c$$

.

Call the resulting net of projectors $\mathcal{N}(G)$. Call the edges of G the *external edges* of G. Call the edges of $\mathcal{N}(G)$ the *internal edges* of G. It is easy to see that admissible loops on $\mathcal{N}(G)$ correspond to loops on G that traverse an edge no move than once and either have no edges or two edges in common with a given vertex. In particular, the region boundaries are such loops and all others may be obtained by mod-2 addition of these cycles.

Accordingly, there is a set, $\Lambda(G)$, of *admissible cycles* on the graph G. For a given labelling of G by integers there will then be various possibilities for realizing these cycles as loops in specific states (permutation choices at the bars) of $\mathcal{N}(G)$. (For example, in the case of the theta graph labelled a, b, c we have three cycles and each cycle carries, respectively, x, y or z loops with $x = (a+c-b)/2, y = (a+b-c)/2, z = (b+c-a)/2$. In this case the number of loops carried by a given cycle is determined by the labelling of the graph.) In the case of an arbitrary trivalent plane graph G, the total number of loops can vary, and so we consider a collection of possible loop configurations. A *loop configuration* $\mathcal{L}$ consists in a non-empty subset $\mathcal{L} \subset \Lambda(G)$ together with positive integer labels on each cycle in $\mathcal{L}$ such that for each edge of G the label of that edge is equal to the sum of the labels on cycles in $\mathcal{L}$ incident to the edge of G.

Thus, associated to a loop configuration $\mathcal{L} \subset \Lambda(G)$ there is a multiplicity of loops obtained by taking $a(l)$ copies of l for each cycle $l \in \mathcal{L}$ with label $a(l)$. Call this the *bundle of loops associated with l.* (In the theta, we had three bundles with x, y and z loops respectively.) Given a loop configuration $\mathcal{L}$, let $P_{\mathcal{L}}(N)$ denote the number of ways to color the loops of $\mathcal{L}$ with N colors so the loops that share an edge receive different colors. (In the case of theta, $P_{\mathcal{L}}(N) = (N)(N-1)\cdots(N-(x+y+z)+1)$.) Let $\mathcal{L}!$ denote the product of the factorials of the loop multiplicities $a(l)$ running over all the cycles in $\mathcal{L} : \mathcal{L}! = \prod_{l \in \mathcal{L}} a(l)!$.

Let $\mathcal{E}!$ denote the product of the factorials of the external edge labels of G. Let $\mathcal{I}!$ denote the product of the factorials of the internal edge labels of G.

Theorem 4 *Let G be a trivalent plane graph with admissible edge labels and associated projector network $\mathcal{N}(G)$. Let $||G||$ denote the positive N evaluation of $\mathcal{N}(G)$: $||G|| = ||\mathcal{N}(G)||(N)$. Then*

$$||G|| = \frac{\mathcal{I}!}{\mathcal{E}!} \sum_{\mathcal{L}} P_{\mathcal{L}}(N)/\mathcal{L}!$$

where $\mathcal{L}$ runs over all loop configurations of G, and $\mathcal{I}!, \mathcal{E}!$ and $\mathcal{L}!$ are as defined just prior to the statement of the theorem.

Proof: Consider a bar in the network $\mathcal{N}(G)$. If $\mathcal{L}$ is a loop configuration for G, then $\mathcal{L}$ specifies everything about a particular loop state except the specific permutations at each bar that effect the connection of loops in the pattern of $\mathcal{L}$. Thus, associated with $\mathcal{L}$ there is a specific collection of loop states $S(\mathcal{L})$ and $\mathcal{L}'s$ contribution to $||G||$ is therefore $|S(\mathcal{L})|P_{\mathcal{L}}(N)/\mathcal{E}!$ since (by the Jordan curve theorem) all states contribute the same sign, each bar is divided by the factorial of its corresponding external edge in the definition of $||G||$, and $|S(\mathcal{L})|$ denotes the cardinality of $S(\mathcal{L})$. Thus we can write $||G|| = \frac{1}{\mathcal{E}!} \sum_{\mathcal{L}} |S(\mathcal{L})| P_{\mathcal{L}}(N)$.

It remains to prove that the cardinality of $S(\mathcal{L})$ is $\mathcal{I}!/\mathcal{L}!$. In order to see this, take note of the following observations:

1) Given a state S in $S(\mathcal{L})$ it is possible to obtain any other state S' in $S(\mathcal{L})$ from S by changing permutations at some or all of the bars in $\mathcal{N}(G)$.

2) Since a given (oriented) cycle in $\mathcal{L}$ enters a bar along one internal edge and leaves by another internal edge, any permutation applied to a bar in changing S to S' must only act to permute strands within the internal edges incident to that bar.

3) For purposes of bookkeeping, orient all the internal edges of $G(\mathcal{N}(G))$ so that each internal edge can be said to "leave" one bar and "enter" another. Call these bars respectively the "source" and the "target" of that edge. In the state S the bundle of lines forming a given internal edge e is partitioned into subsets $S_1(e), \ldots, S_k(e)$ corresponding to the distinct cycles of $\mathcal{L}$ that are incident to e.

4) Let σ be a permutation on l objects where l is the number of strands on a given internal edge of G. Let S be a given state in $\mathcal{S}(\mathcal{L})$. Then the source and target bars for e are each equipped with given permutations. Modify the source bar for e with σ by multiplying it on the output side by σ (extended to identity on the remaining strands).

   Now, regarding σ as a product of transpositions, note that each transposition either permutes one of the subsets $S_j(e)$ (see 3) above) or it interchanges strands between two of these subsets $S_j(e), S_k(e)$ with $j \neq k$. Call the transposition *internal* if it permutes within a subset and *external* otherwise. Let $\hat{\sigma}(e)$ denote the product of the external transpositions in σ. Modify the target bar of e on its receiving side by the inverse of $\hat{\sigma}(e)$. Call the new state obtained S^σ.

   Note that S^σ preserves the *form* of $\mathcal{L}$, but may include permutations on the bundles of strands corresponding to given cycles in $\mathcal{L}$. States of this type form a set $\hat{\mathcal{S}}(\mathcal{L})$ and it is easy to see that the cardinality of $\hat{\mathcal{S}}(\mathcal{L})$ is related to the cardinality of $\mathcal{S}(\mathcal{L})$ by the formula $|\hat{\mathcal{S}}(\mathcal{L})| = \mathcal{L}!|\mathcal{S}(\mathcal{L})|$. (In order for a bundle of strands in a loop to have no self crossing cycles [and hence give part of a state in $\mathcal{S}(\mathcal{L})$] it is necessary and sufficient that the product of the permutations received at the bars be the identity. This restriction gives the formula above.)

5) Finally, we claim that $\hat{\mathcal{S}}(\mathcal{L})$ has cardinality $\mathcal{I}!$, i.e., $|\hat{\mathcal{S}}(\mathcal{L})| = \mathcal{I}!$. This follows from the construction described in 4): starting with any element $S \in \hat{\mathcal{S}}(\mathcal{L})$ and permutations for each internal edge of G, we obtain a new element S' of $\hat{\mathcal{S}}(\mathcal{L})$. Any element S' can be reached this way from S *and* it is easy to see *given S and any other element S'*

how to reconstruct the permutations for each edge (this being a local process). Uniqueness of the association $\sigma, \hat{\sigma}(e)$ is guaranteed by using a standard minimal representation of the permutations as products of transpositions.

With these observations in hand, we have

$$|\mathcal{S}(\mathcal{L})| = \frac{|\hat{\mathcal{S}}(\mathcal{L})|}{\mathcal{L}!} = \frac{\mathcal{I}!}{\mathcal{L}!}.$$

This completes the proof of the formula

$$\|G\| = \frac{\mathcal{I}!}{\mathcal{E}!} \sum_{\mathcal{L}} \mathcal{P}_{\mathcal{L}}(N)/\mathcal{L}!$$

■

Remark 12 *The following figure will clarify the argument in the proof of Theorem 4.*

(A)

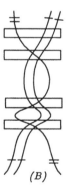

(B)

In (A) two groups of "cables" enter the lower bar and exit from the top bar. The permutation σ is applied to the lower bar in (B). We match σ with $\hat{\sigma}$ (applied just before the top bar). $\hat{\sigma}$ is a standard connection to re-direct the cables as groups of lines. In general the result of placing σ and $\hat{\sigma}$ may cause internal permutations on the cables, but $\hat{\sigma}$ is chosen to be the identity on internal cables and hence the internal permutations all arise from choices in σ.

8.5   The Tetrahedron

We now apply Theorem 4 to the case of the tetrahedral network
$$T \begin{bmatrix} a & b & e \\ c & d & f \end{bmatrix} = G$$

There are seven admissible cycles on the tetrahedral graph G. These cycles are illustrated in Figure 8 and Figure 9. In the first figure we show the cycles on the graph G. In the second they are drawn on the associated projector nerwork $\mathcal{N}(G)$.

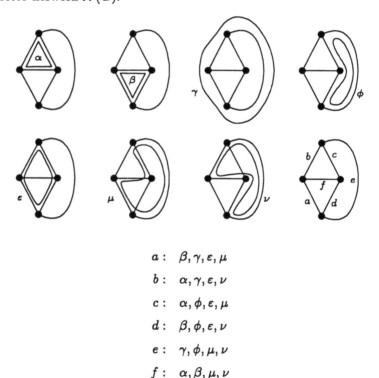

$$
\begin{aligned}
a &: \quad \beta, \gamma, \varepsilon, \mu \\
b &: \quad \alpha, \gamma, \varepsilon, \nu \\
c &: \quad \alpha, \phi, \varepsilon, \mu \\
d &: \quad \beta, \phi, \varepsilon, \nu \\
e &: \quad \gamma, \phi, \mu, \nu \\
f &: \quad \alpha, \beta, \mu, \nu
\end{aligned}
$$

Figure 8: Seven Cycles on the Tetrahedron

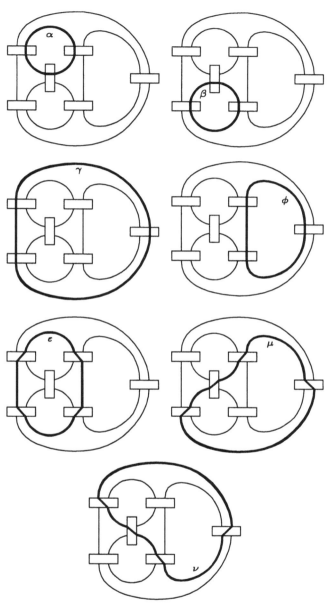

Figure 9: Seven Basic Cycles on the Tetrahedron Net

The edge labels on the tetrahedral graph are a, b, c, d, e, f. The cycles are labelled $\alpha, \beta, \gamma, \phi, \varepsilon, \mu, \nu$ with the first four traversing three edges of G while the last four traverse four edges of G. Letting $\alpha, \beta, \gamma, \phi, \varepsilon, \mu, \nu$ denote the number of individual loops carried by a specific loop configuration $\mathcal{L}$ of $\mathcal{N}(G)$, we see from figure 8 that the following equations hold.

$$
\begin{aligned}
a &= \beta + \gamma + \varepsilon + \mu \\
b &= \alpha + \gamma + \varepsilon + \nu \\
c &= \alpha + \phi + \varepsilon + \mu \\
d &= \beta + \phi + \varepsilon + \nu \\
e &= \gamma + \phi + \mu + \nu \\
f &= \alpha + \beta + \mu + \nu.
\end{aligned}
$$

Let

$$
S = \alpha + \beta + \gamma + \phi + \varepsilon + \mu + \nu.
$$

Let

$$
\begin{aligned}
a_1 &= (a + b + f)/2 = S - \phi \\
a_2 &= (b + c + e)/2 = S - \beta \\
a_3 &= (c + d + f)/2 = S - \gamma \\
a_4 &= (a + d + e)/2 = S - \alpha.
\end{aligned}
$$

(Every 3-cycle goes through all but one vertex.) Let also

$$
\begin{aligned}
b_1 &= (b + d + e + f)/2 = S + \nu \\
b_2 &= (a + c + e + f)/2 = S + \mu \\
b_3 &= (a + b + c + d)/2 = S + \varepsilon
\end{aligned}
$$

(Every 4-cycle misses two edges.)

Thus we have $a_i \le S \le b_j$ for $i = 1, \ldots, 4$ and $j = 1, \ldots, 3$. Each value of S in this range gives non-negative solutions for $\alpha, \beta, \gamma, \phi, \varepsilon, \mu, \nu$ and hence a specific loop configuration $\mathcal{L}(S)$.

Note that the set of twelve values $\{ b_j - a_i | i = 1, \ldots, 4; j = 1, \ldots, 3 \}$ is exactly the set of values of the internal edges of G. (For example $b_3 - a_2 = (a + e - d)/2$ an internal edge value at the aed-vertex of G.) Thus we have

$$
\begin{aligned}
\mathcal{L}! &= \mathcal{L}(S)! = \alpha! \beta! \gamma! \phi! \varepsilon! \mu! \nu! \\
\mathcal{E}! &= a! b! c! d! e! f! \\
\mathcal{I}! &= \textstyle\prod_{i,j} (b_i - a_j)!
\end{aligned}
$$

Note that

$$
\begin{aligned}
\alpha &= a_4 + S, & \varepsilon &= b_3 - S \\
\beta &= a_2 + S & \mu &= b_2 - S \\
\gamma &= a_3 + S & \nu &= b_1 - S. \\
\phi &= a_1 + S
\end{aligned}
$$

Hence $\mathcal{L}! = \prod_{i=1}^{4}(a_i + S)! \prod_{j=1}^{3}(b_j - S)!$.

Finally, in each loop configuration all loops are mutually touching. Therefore, $\mathcal{P}_{\mathcal{L}}(N) = (N(N-1)\cdots(N-S+1)$ since in $\mathcal{L}(S)$ there are $S = \alpha + \beta + \gamma + \phi + \varepsilon + \mu + \nu$ loops all touching each other. Therefore, by Theorem 4 we have shown

Proposition 12 *Let G denote the tetrahedral net with edge labels $a, b, c, d, e,$ f. Let $a_1, \ldots, a_4$ and $b_1, \ldots, b_3$ be defined as above. Then the positive N evaluation of the net G is given by the formula*

$$\|G\|(N) =$$

$$= \frac{\prod_{\substack{i=1,\ldots,3 \\ j=1,\ldots,4}} (b_i - a_j)!}{a!b!c!d!e!f!} \sum_{max\{a_i\} \leq S \leq min\{b_j\}} \frac{N(N-1)\cdots(N-S+1)}{\prod_{i=1}^{4}(a_i + S)! \prod_{j=1}^{3}(b_j - S)!}.$$

Proof: $\|G\| = \frac{\mathcal{I}!}{\mathcal{E}!} \sum_{\mathcal{L}} \mathcal{P}_{\mathcal{L}}(N)/\mathcal{L}!$. ∎

Specializing N to -2 we obtain the value of the $\|G\|(-2) = \langle G \rangle$ $(A = -1)$.

$$\langle G \rangle (A = -1) =$$

$$= \frac{\prod_{i,j}(b_i - a_j)!}{a!b!c!d!e!f!} \sum_{max\{a_i\} \leq S \leq min\{b_j\}} \frac{(-1)^S(S+1)!}{\prod_i(a_i + S)! \prod_j(b_j - S)!}.$$

Replacing $n!$ by $[n]!$ with $[n] = (A^{-2n} - A^{-2})/(A^2 - A^{-2})$ gives a formula, for general A of $\langle G \rangle (A)$.

This general formula has been verified directly by using an induction argument and the Jones-Wenzl recursion relation in *[MV92]*. We will omit this induction argument from our exposition.

Discussion. The principle of replacing integers by quantum integers has a wider application in the context of quantum groups. Here we have presented it as a heuristic method for guessing a general formula, that can then be verified by rigorous induction. The subject obviously needs more exploration. In particular it is not yet clear how to apply these heuristics to arbitrary plane trivalent graphs G. Theorem 4 gives a rigorous formula for $\|G\|(N)$ and hence for $\langle G \rangle (A = -1) = \|G\|(-2)$, but it is not clear how to "quantize" $\mathcal{P}_{\mathcal{L}}(-2)$ in the general case. If we knew the answer to this quantization – call it $[\mathcal{P}_\mathcal{L}]$ – then we could write

$$< G > (A) = \frac{[\mathcal{I}]!}{[\mathcal{E}]!} \sum_\mathcal{L} \frac{[\mathcal{P}_\mathcal{L}(-2)]}{[\mathcal{L}]!}$$

where all the other quantizations in the formula are standard quantum integers. This would give a chromatic formula for the evaluation of arbitrary nets.

In fact, it follows from our recoupling theory that it is sufficient to know the theta and tetrahedral evaluations, since any trivalent graph can be simplified by successive recoupling. Nevertheless, it would be a esthetically pleasing to have a full generalization of the chromatic formula.

Remark 13 *The case $A = -1$ corresponds to standard angular momentum theory. For example, our formula for the tetrahedral net it $A = -1$ is a multiple of a standard Racah coefficient (see [BL79]) and the orthogonality and Biedenhan-Elliot identities are fundamental to the study of angular momentum and $SU(2)$. The idea behind the chromatic method explained here first appears in [Pen71].*

Chapter 9

A Summary of Recoupling Theory

The purpose of this Chapter is to summarize the recoupling theory that we have developed over the first eight chapters of our monograph. It is intended that a reader interested primarily in the applications could read the introduction, Chapters 1, 2, 3 plus this summary in order to work with the remaining chapters. Nevertheless, here our style is telegraphic. We present formulas and diagrams out of context, but with some references to their sources earlier in the paper.

9.1   Bracket Polynomial

$$\left\langle \asymp \right\rangle = A \left\langle \smile\!\!\!\frown \right\rangle + A^{-1} \left\langle \,)(\, \right\rangle$$

$$d = \text{loop value} = \left\langle \bigcirc \right\rangle = -A^2 - A^{-2}.$$

Abbreviation: $\asymp = A \,\smile\!\!\!\frown\, + A^{-1}\,)(\,$, $\bigcirc = -A^2 - A^{-2}.$

9.2   Temperley-Lieb Algebra T_n

$$U_i^2 = dU_i$$

$$U_iU_{i\pm1}U_i = U_i$$

$$U_iU_j = U_jU_i \ , \ |i-j| \geq 2.$$

9.3   Chebyschev Polynomials

$$\Delta_{-1} = 0, \ \Delta_0 = 1$$

$$\Delta_{n+1} = d\Delta_n - \Delta_{n-1}.$$

If $\ d \ = \ x + x^{-1}\ $ then

$$\Delta_n = (x^{n+1} - x^{-n-1})/(x - x^{-1}).$$

If $\ d \ = \ -A^2 - A^{-2}\ $ then

$$\Delta_n = (-1)^n \frac{A^{2n+2} - A^{-2n-2}}{A^2 - A^{-2}}.$$

If $\ A \ = \ e^{i\pi/2r}$

$$q = A^2 = e^{i\pi/r}$$

$$\Delta_n = (-1)^n \sin((n+1)\pi/r)/\sin(\pi/r).$$

$$\implies \left\{ \begin{array}{l} \Delta_n \neq 0 \quad \text{for} \quad 0 \leq n \leq r-2 \\ \Delta_{r-1} = 0 \end{array} \right\}.$$

9.4 Quantum Integers

$$\{n\} = \frac{1 - A^{-4n}}{1 - A^{-4}}$$

$$\{n\}! = \prod_{k=1}^{n}\{n\} = \sum_{\sigma \in S_n}(A^{-4})^{t(\sigma)}$$

$$S_n = \text{permutations on } n \text{ letters.}$$

$$t(\sigma) = \text{least number of transpositions in a}$$
$$\text{factorization of } \sigma \in S_n.$$

$$[n] = \frac{A^{2n} - A^{-2n}}{A^2 - A^{-2}} = (-1)^{n-1}\Delta_{n-1}$$

Note: $[n](A = \pm 1) = n$.

9.5 q-Symmetrizer

$(q = A^2)$

$$\begin{array}{c}|n\\ \boxed{}\\ |\end{array} = \frac{1}{\{n\}!}\sum_{\sigma \in S_n}(A^{-3})^{t(\sigma)}\begin{array}{c}|n\\ \boxed{\hat{\sigma}}\\ |\end{array}$$

$\hat{\sigma}$ = minimal positive ($\asymp$) braid lift of permutation diagram for σ.

$$\left[\text{Evaluations taken modulo } \asymp = A\,\smile\!\!\frown + A^{-1}\right)\,\Big(\cdot\Big]$$

Facts:

$$\begin{array}{c}|n\\ \boxed{}\\ |n\\ \boxed{}\\ |\end{array} = \begin{array}{c}|n\\ \boxed{}\\ |\end{array}$$

$$n\,\bigcirc\!\!\!\!\boxed{} = \Delta_n = (-1)^n\frac{A^{2n+2} - A^{-2n-2}}{A^2 - A^{-2}}.$$

$$\begin{array}{c}|n\\ \boxed{}\\ \boxed{U_i}\\ |\end{array} = 0 \quad \text{for} \quad 1 \le i \le (n-1)$$

9.6   Jones-Wenzl Projectors

(Inductive Construction for q-Symmetrizers)

$$\boxed{}^{n} \quad \in \quad T_n$$

$$\begin{array}{c} \text{diagram} \end{array} \;=\; \begin{array}{c} \text{diagram} \end{array} \;-\; \frac{\Delta_n}{\Delta_{n+1}} \begin{array}{c} \text{diagram} \end{array}$$

$$\boxed{}^{1} \;=\; \Big|^{1}$$

(n denotes a parallel cable of n strands. An unlabelled strand is a single strand.)

9.7   Curl and Projector

$$\begin{array}{c} \text{diagram} \end{array} \;=\; (-1)^n A^{-n(n+2)} \begin{array}{c} \text{diagram} \end{array}^{n} \qquad \text{(Prop. 6)}$$

9.8   Loop and Projector

$$\begin{array}{c} \text{diagram} \end{array} \;=\; (-A^{2n+2} - A^{-2n-2}) \begin{array}{c} \text{diagram} \end{array}^{n} \qquad \text{(Prop. 7)}$$

9.9   3-Vertex

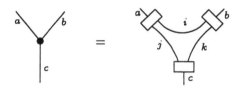

$$i = (a + b - c)/2$$
$$j = (a + c - b)/2$$
$$k = (b + c - a)/2$$

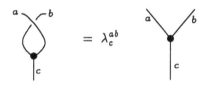

$$\lambda_c^{ab} = (-1)^{(a+b-c)/2} A^{(a'+b'-c')/2} \,, \quad \text{where} \quad x' = x(x+2)$$

9.10 θ-Net

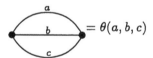

(a, b, c) *admissible* means $a + b - c, a + c - b, b + c - a$ all positive and even. If $q = e^{i\pi/r}$ then (a, b, c) is *q-admissible* if it is admissible and $a + b + c \leq 2r - 4$.

(i)

$$\theta(a, b, c) = \frac{(-1)^{m+n+p}[m + n + p + 1]![n]![m]![p]!}{[m + n]![n + p]![p + m]!} \,, \text{where}$$

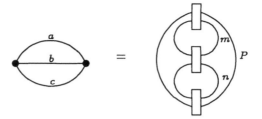

(ii) $\theta(a, b, c) \neq 0 \Leftrightarrow (a, b, c)$ is q-admissible (when $A^2 = q = e^{i\pi/r}$)

(iii) $\qquad = \dfrac{\theta(a, b, c)}{\Delta_a} \quad \boxed{} \quad \delta_{aa'}$

9.11   Tetrahedral Net

$$\left.\begin{array}{c} \end{array}\right)\, E \;=\; Tet \begin{bmatrix} A & B & E \\ C & D & F \end{bmatrix}$$

$$Tet \begin{bmatrix} A & B & E \\ C & D & F \end{bmatrix} = \frac{\mathcal{I}!}{\mathcal{E}!} \sum_{m \leq s \leq M} \frac{(-1)^s [s+1]!}{\prod_i [s - a_i]! \prod_j [b_j - s]!}$$

$$
\begin{aligned}
\mathcal{I}! &= \textstyle\prod_{i,j} [b_j - a_i]! \\
\mathcal{E}! &= [A]![B]![C]![D]![E]![F]! \\
a_1 &= \tfrac{1}{2}(A + D + E) & b_1 &= \tfrac{1}{2}(B + D + E + F) \\
a_2 &= \tfrac{1}{2}(B + C + E) & b_2 &= \tfrac{1}{2}(A + C + E + F) \\
a_3 &= \tfrac{1}{2}(A + B + F) & b_3 &= \tfrac{1}{2}(A + B + C + D) \\
a_4 &= \tfrac{1}{2}(C + D + F) & m &= \max\{a_i\} \quad M = \min\{b_j\}
\end{aligned}
$$

9.12 q-$6j$ Symbols

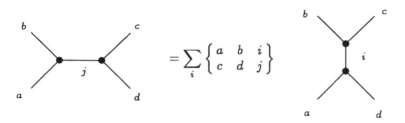

$$= \sum_i \left\{ \begin{matrix} a & b & i \\ c & d & j \end{matrix} \right\}$$

Sum over admissibles for generic q. Sum over q-admissibles for q a $2r$-th root of 1.

$$\left\{ \begin{matrix} a & b & i \\ c & d & j \end{matrix} \right\} = \frac{Tet \begin{bmatrix} a & b & i \\ c & d & j \end{bmatrix} \Delta_i}{\theta(a,d,i)\theta(b,c,i)}$$

9.13 Orthogonality Identity

$$\sum_i \left\{ \begin{matrix} a & b & i \\ c & d & j \end{matrix} \right\} \left\{ \begin{matrix} d & a & k \\ b & c & i \end{matrix} \right\} = \delta_{kj}$$

9.14 Biedenharn-Elliot (Pentagon) Identity

$$\sum_{m=0}^{r-1} \left\{ \begin{matrix} a & i & m \\ d & e & j \end{matrix} \right\} \left\{ \begin{matrix} b & c & l \\ d & m & i \end{matrix} \right\} \left\{ \begin{matrix} b & l & k \\ e & a & m \end{matrix} \right\} = \left\{ \begin{matrix} b & c & k \\ j & a & i \end{matrix} \right\} \left\{ \begin{matrix} k & c & l \\ d & e & j \end{matrix} \right\}.$$

9.15 Two Special Cases

$$\bigg) \bigg(= \sum_i \frac{\Delta_i}{\theta(a,b,i)} \quad \bigvee_{i}$$

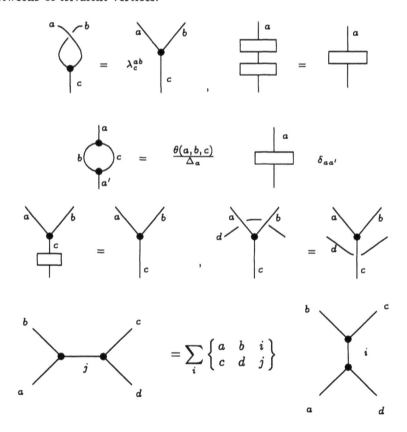

$$\times = \sum_i \frac{\Delta_i}{\theta(a,b,i)} \lambda_i^{ab}$$

(see Sec. 9.9 for λ_i^{ab}.)

9.16   Axiomatics

The following properties suffice to deduce (almost) everything else about networks of trivalent vertices.

We have ignored admissibility hypotheses, and one may wish to add the formulas from Sections 7 and 8. It is not hard to formalize the axiomatics. Here we just point out salient features of the theory as constructed from q-symmetrizers and Temperley-Lieb algebra.

Chapter 10

A 3-Manifold Invariant by State Summation

In this Chapter we explain how to apply the Temperley-Lieb recoupling theory of Section 4 to produce an invariant of 3-manifolds. This invariant is identical, up to a normalization, with the state summation invariant of Turaev and Viro [TV92], as was later proved by Piunikhin [Piu92]. Their invariant is based upon the re-coupling theory of the quantum group $SL(2)q$ for q a root of unity. Our tangle-theoretic recoupling theory provides an alternative approach to these constructions. The material in this Chapter and in the preceding ones provides a self-contained account of the construction of an invariant of 3-manifolds via state summation.

10.1 Matveev-Piergallini Moves

This approach to an invariant of 3-manifolds is based on the Alexander moves [Ale30] as reformulated by Turaev and Viro [TV92] in their exposition of the Matveev [Mat88] and Piergallini [Pie88] moves for special spines for 3-manifolds. We begin with a short summary about special spines and the Matveev-Piergallini moves. First, refer to Figure 10.

In this figure it is illustrated how six two-dimensional cells, sharing a single vertex, are dual to a tetrahedron in the sense that the vertex is an interior point of the tetrahedron, each one-cell of the tetrahedron pierces an interior point of a corresponding two-cell. The complex formed by these two-cells has edges with three two-cells sharing each edge. These edges pierce the faces of the tetrahedron.

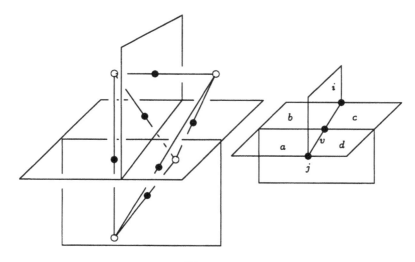

Figure 10

A triangulated 3-manifold gives rise to a global dual complex of this type, with one or more 3-cells, each homeomorphic to a ball. Such a cell complex, will be called a *special spine* of the 3-manifold. In [TV92] Turaev and Viro recount how Alexander's moves for triangulations can be re-cast into a set of moves on special spines. These Matveev-Piergallini moves are illustrated in Figure 11. There are three basic moves, labelled II,III and B. The move of type II involves the creation or removal of two vertices. The move of type III passes from a local situation with two vertices to a local situation with three vertices. The move B (bubble move) creates or destroys a three-cell. We have indicated two versions of bubble moves, one isolated, one involving an edge. In the presence of moves II and III, these bubble moves are equivalent (and the edge-version will be more convenient for our use). In Section 13.4 we provide a new proof of the sufficiency of these moves for constructing homeomorphisms of 3-manifolds.

Now recall the formalism of the recoupling theory of the last section. Given a coloring of the 2-cells of special spine for $\mathcal{M}$, we assign *weights* to the vertices, edges and faces of the special spine as follows: The weight of a vertex v, denoted $TET(v|\sigma)$ is the value of the tetrahedral net associated with v. Thus for v as Figure 10, we have

(σ denotes the coloring of the edges of the net) where this net is evaluated by the technique explained in Chapter 8.

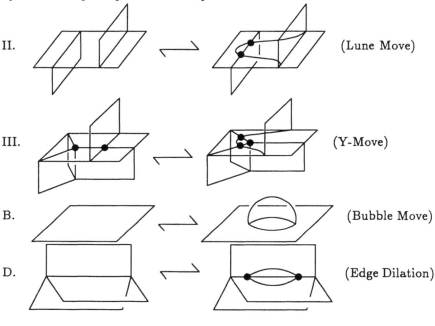

II.   (Lune Move)

III.   (Y-Move)

B.   (Bubble Move)

D.   (Edge Dilation)

Figure 11

10.2   A Partition Function

The above transformations on special spines are called *Matveev-Piergallini moves* [Mat88], [Pie88]. The weight of an edge is the value of the θ-net associated with it, where the lines of the θ are labelled with the colors of the three 2-cells incident to the edge. Thus if e denotes an edge with incident colors a, b, c, then $\theta(a, b, c) =$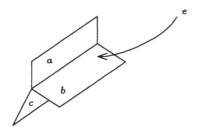

Finally, the weight of a colored face is the quantum integer Δ_i associated with that face colored i.

With these definitions of weights, we define a *partition function*

$$TV_{\mathcal{M}^3} = TV_{\mathcal{M}^3, r} \quad \text{associated with} \quad \mathcal{M}^3$$

and a given finite color set $\{0, 1, 2, \ldots, r - 2\}$. $TV_{\mathcal{M}^3}$ is defined by the formula:

Definition 7

$$TV_{\mathcal{M}^3} = \sum_\sigma \prod_{v, e, f} \frac{TET(v|\sigma) \Delta_{\sigma(f)}^{\chi(f)}}{\theta(e|\sigma)^{\chi(e)}}$$

where $\chi(f)$ is the Euler characteristic of the 2-cell f in the spine of $\mathcal{M}^3$, and $\chi(e)$ is the Euler characteristic of the edge. That is, $\chi(e) = 1$ if e contains graphical nodes, and $\chi(e) = 0$ if e is a loop without nodes.

In order to see that this partition function has the required invariance properties, we now discuss homeomorphisms of the 3-manifold with respect to the special spine in relation to this partition function. For this purpose it is useful to adopt a 2-dimensional notation (a shadow world notation as in [KR88]) for local colorings of the special spine. Refer to *Figure 11*. We shall abbreviate this equivalence by the flat diagram

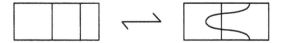

Here it is understood that the outer edges of the diagram are simply a frame for this snapshot of the 2-cells in the spine of a 3-manifold $\mathcal{M}$. The 2-cell extending above and below the horizontal 2-cell in the left part of Figure 11 are here abbreviated to the two vertical lines in the interior of the left-hand flat diagram. Thus we have the correspondence shown below:

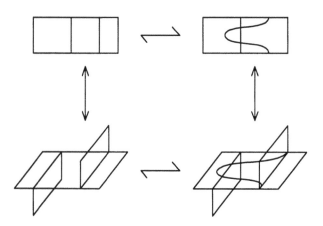

In the flat diagrams, interior line segments and curves correspond to 2-cells that abut perpendiculary to the flat planes either from above or below. An intersection of two curves in the flat diagram corresponds to a vertex in the spine. At such an intersection, four of the local cells adjacent to the vestex are indicated by cells in the flat diagram. The other two cells are indicated by the two segments that cross at the flat vertex. Thus, if these cells are colored a, b, c, d, i and j then we shall indicate this on the flat diagrams by coloring 4 cells and two segments, as shown below.

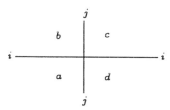

Segments retain the same color across a crossing, since they correspond to the intersection of a 2-cell with the plane (a, b, c, d) from above or below.

Now refer to Figure 11, part III.

By the same conventions, this is abbreviated by the flat diagrams

Here we see that a vertex in the spine may *also* be represented in a flat diagram by a trivalent graphical vertex ⟩●—. Each edge of the trivalent vertex correspond to a distinct 2-cell, and the three cells in the flat diagram at this vertex represent the remaing three 2-cells in the spine at this site.

Finally, the abstract bubble move B of Figure 10 is represented in flat diagrams by

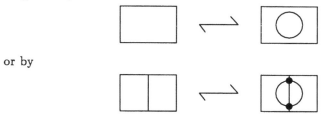

or by

We shall use the notation of flat diagrams to aid the analysis of this particular function's behaviour under the Matveev-Piergallini moves.

Nota Bene. With the exception of the bubble move verification, we shall assume that $\chi(e) = 1$ for edges in the demonstrations. Remaining cases are left for the reader.

10.3   Invariance under Lune Move

First consider move II (Figure 11) in flat notation, with colors added as shown below:

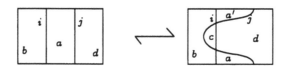

In order for the partition function TV_{M^3} to be invariant under this move it will suffice to show that (i) the right hand side necessarily vanishes when $a \neq a'$ and (ii) for fixed colors a, b, d, i and j the restriction of the state sum on the left (to these states with these fixed colors) is equal to the state sum on the right hand side where these colors are still fixed $(a' = a)$ and c runs over all admissible possibilities in the color set. Let $\mathcal{R}$ denote this restriction of the partition function for the right-hand diagram. Then $\mathcal{R}$ is given by the formula shown below.

$$\mathcal{R} = \mathcal{R}_0 \frac{\triangle_a^{\chi_a} \triangle_{a'}^{\chi_{a'}} \triangle_b^{\chi_b} \triangle_d^{\chi_d} \triangle_i^{\chi_i} \triangle_j^{\chi_j}}{\theta(a,b,i)\theta(a,d,j)}$$

$$\sum_c \frac{\displaystyle \left(\!\!\begin{array}{c}b \;/\!\!\!\backslash\; c\\ \triangle\\ a\;\backslash\!\!\!/\;d\end{array}\!\!\right)_{i}^{\,j} \left(\!\!\begin{array}{c}a' /\!\!\!\backslash\; d\\ \triangle\\ b\;\backslash\!\!\!/\;c\end{array}\!\!\right)_{i}^{\,j} \triangle_c^{\chi_c}}{\theta(b,c,j,)\theta(c,d,i)\theta(a',i,b)\theta(a',d,j)}$$

$$= \mathcal{R}_0 \frac{\triangle_a^{\chi_a} \triangle_{a'}^{(\chi_a - 1)} \triangle_b^{\chi_b} \triangle_d^{\chi_d} \triangle_i^{\chi_i} \triangle_j^{\chi_j}}{\theta(a,b,i)\theta(a,d,j)}$$

$$\sum_c \left[\frac{\displaystyle \left(\!\!\begin{array}{c}j \;/\!\!\!\backslash\; d\\ \triangle\\ b\;\backslash\!\!\!/\;i\end{array}\!\!\right)_a^{\,c} \triangle_c}{\theta(b,c,j)\theta(c,d,i)}\right]\left[\frac{\displaystyle \left(\!\!\begin{array}{c}d \;/\!\!\!\backslash\; j\\ \triangle\\ i\;\backslash\!\!\!/\;b\end{array}\!\!\right)_c^{\,a'} \triangle_{a'}}{\theta(a',i,b)\theta(a',d,j)}\right] .$$

Here $\mathcal{R}_0$ denotes the contribution to the partition function from the rest of the spine. Note that $\chi_c = 1$, since the region labelled c is homeomorphic to a disk. Applying Proposition 7 to the last formula, we have

$$\mathcal{R} = \frac{\mathcal{R}_0 \Delta_a^{\chi_a} \Delta_{a'}^{\chi_{a'}-1} \Delta_b^{\chi_b} \Delta_d^{\chi_d} \Delta_i^{\chi_i} \Delta_j^{\chi_j}}{\theta(a,b,i)\theta(a,d,j)} \sum_c \left\{ \begin{matrix} d & j & c \\ b & i & a \end{matrix} \right\} \left\{ \begin{matrix} i & d & a' \\ j & b & c \end{matrix} \right\}.$$

By Proposition 9, we have that

$$\sum_c \left\{ \begin{matrix} d & j & c \\ b & i & a \end{matrix} \right\} \left\{ \begin{matrix} i & d & a' \\ j & b & c \end{matrix} \right\} = \left\{ \begin{matrix} 1 & \text{if} & a = a' \\ 0 & \text{otherwise.} \end{matrix} \right.$$

Hence $\mathcal{R} = 0$ if $a \neq a'$, as desired; and *if* $a = a'$ then

$$\mathcal{R} = \mathcal{R}_0 \frac{\Delta_a^{(\chi_a + \chi_{a'} - 1)} \Delta_b^{\chi_b} \Delta_d^{\chi_d} \Delta_i^{\chi_i} \Delta_j^{\chi_j}}{\theta(a,b,i)\theta(a,d,j)}.$$

Since the region labelled a in the left hand side of the flat diagrams for move II has Euler characteristic $\chi_a + \chi_{a'} - 1$, it follows at once that $\mathcal{R}$ is equal to the contribution from the left-hand side. This proves invariance under move II, the lune move.

10.4   Invariance under the Y-move

We now turn to the the behaviour of the partition function under the type III move. In order to facilitate this calculation, it is useful to note the labelled correspondence between the vertex as a flat trivalent diagram and the vertex as a flat four-valent diagram. The correspondence is shown below – with the associated pictures of the spine and the associated tetrahedral net.

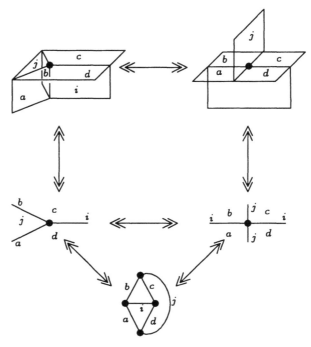

Now recall the statement of the pentagon (Biedenharn-Elliot) identity from Proposition 10:

$$\sum_m \begin{Bmatrix} a & i & m \\ d & e & j \end{Bmatrix} \begin{Bmatrix} b & c & l \\ d & m & i \end{Bmatrix} \begin{Bmatrix} b & l & k \\ e & a & m \end{Bmatrix} = \begin{Bmatrix} b & c & k \\ j & a & i \end{Bmatrix} \begin{Bmatrix} k & c & l \\ d & e & j \end{Bmatrix}.$$

Here the $q-6j$ symbols are taken at a root of unity, and the m runs through the corresponding set of values in the color set, with all triples q-admissible.

Compare this equation with the labelled diagram of the type III move, the Y-move shown below:

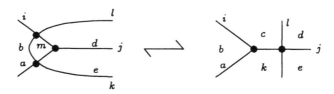

Let $\mathcal{L}$ denote the partial sum for the partition function for the left-hand diagram (as part of the partition function for a 3-manifold with corresponding special spine, and all colors but m held fixed). Then

$$\mathcal{L} = \mathcal{L}_0 \sum_m \frac{\Delta_i^{\chi_i} \Delta_a^{\chi_a} \Delta_j^{\chi_j} \Delta_b^{\chi_b} \Delta_c^{\chi_c} \Delta_d^{\chi_d} \Delta_e^{\chi_e} \Delta_k^{\chi_k} \Delta_l^{\chi_l} \Delta_m^1}{\Theta}$$

$$\times \quad \begin{array}{c} \text{(diagram)} \end{array} \quad \begin{array}{c} \text{(diagram)} \end{array} \quad \begin{array}{c} \text{(diagram)} \end{array} \quad ,$$

where,

$$\Theta = \theta(b,i,c)\theta(b,a,k)\theta(c,l,d)\theta(d,j,e)\theta(e,l,k)\theta(d,i,m)\theta(e,a,m)\theta(b,l,m).$$

Now note that

$$\begin{array}{c} \text{(diagram)} \end{array} \quad = \quad \begin{array}{c} \text{(diagram)} \end{array}$$

since the cyclic orientations at corresponding vertices are identical. Therefore

$$\mathcal{L} = \mathcal{L}_0 \frac{\Delta_i^{\chi_i} \Delta_a^{\chi_a} \Delta_j^{\chi_j} \Delta_b^{\chi_b} \Delta_c^{\chi_c} \Delta_d^{\chi_d} \Delta_e^{\chi_e} \Delta_k^{\chi_k-1} \Delta_l^{\chi_l-1}}{\theta(b,i,c)\theta(d,j,e)}$$

$$\times \sum_m \left\{ \begin{array}{ccc} a & i & m \\ d & e & j \end{array} \right\} \left\{ \begin{array}{ccc} b & c & l \\ d & m & i \end{array} \right\} \left\{ \begin{array}{ccc} b & l & k \\ e & a & m \end{array} \right\}$$

$$= \mathcal{L}_0 \frac{\Delta_i^{\chi_i} \Delta_a^{\chi_a} \Delta_j^{\chi_j} \Delta_b^{\chi_b} \Delta_c^{\chi_c} \Delta_d^{\chi_d} \Delta_e^{\chi_e} \Delta_k^{\chi_k-1} \Delta_l^{\chi_l-1}}{\theta(b,i,c)\theta(d,j,e)}$$

$$\times \left(\frac{\begin{array}{c} \text{(diagram)} \end{array} \Delta_k}{\theta(c,k,j)\theta(b,a,k)} \right) \left(\frac{\begin{array}{c} \text{(diagram)} \end{array} \Delta_l}{\theta(c,l,d)\theta(k,e,l)} \right).$$

(By the Biedenharn-Elliot identity).

A moment's inspection reveals that this is precisely the partition function for the right-hand side of the move III. This completes the proof that TV_M is invariant under the Y-move.

10.5   Behavior under Bubble Move

Finally, we must examine the behaviour of TV_M under the bubble move B of Figure 11. With the labelling shown below we see that if M' is the result of perfoming a bubble move on M, then $TV_{M'} = (\Delta_a^{-1} \sum_{i,j} \Delta_i \Delta_j) TV_{M'}$.

(The Δ_a^{-1} takes care of the change in Euler characteristic in the cell labelled a.) In fact, we have the following Lemma.

Lemma 19 *Let* $q = e^{i\pi/r}$. *Let* $a, b \in \{0, 1, 2, \dots, r-2\}$. *Then*

$$\Delta_a^{-1} \sum_{i,j} \Delta_i \Delta_j = \Delta_b^{-1} \sum_{i,j} \Delta_i \Delta_j$$

where the left hand sum is taken over q-admissible triples (a, i, j) and the right hand sum is taken over q-admissible triples (b, i, j).

Proof: This is a consequence of the orthogonality identity. That is, apply invariance under the Matveev-Piergallini lune move to the flat diagrams shown below:

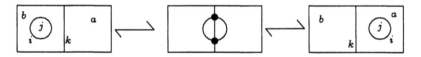

this completes the proof.   ∎

Lemma 20   *For any* $j \in \{0, 1, \dots, r-2\}$, *and* (a, i, j) *q-admissible* $\tau_q = \Delta_a^{-1} \sum_{i,j} \Delta_i \Delta_j = 2r/(q - q^{-1})^2$.

Proof: The previous Lemma shows that the expression on the left is independent of a. Let $a = 0$. Then

$$\Delta_0^{-1} \sum_{i,j} \Delta_{i,j} = \sum_{n=0}^{r-2} \Delta_n \Delta_n = \sum_{n=0}^{r-2} \left(\frac{x^{n+1} - x^{-n-1}}{x - x^{-1}} \right)^2$$

with $x = -q$. Thus

$$\tau_q = \frac{1}{(x - x^{-1})^2} \left(\sum_{n=0}^{r-2} x^{2n+2} + \sum_{n=0}^{r-2} x^{-2n-2} - \sum_{n=0}^{r-2} 2 \right).$$

Since x^2 and x^{-2} are r-th roots of unity, the two first sums above are both equal to -1. Hence

$$\tau_q = \frac{1}{(x - x^{-1})}(-1 - 1 - 2r + 2) = -2r/(q - q^{-1})^2.$$

as stated. ∎

Remark 14 *Observe that this constant is the same as in [TV92].*
It now follows that if the spine of $M^{3'}$ is obtained by a bubble move

$$\left(\text{or edge dilation} \Big/ \quad \longrightarrow \quad \emptyset \right) \text{ from the spine of } M^3, \text{ then}$$

$TV_{M^{3'}} = \tau_q TV_{M^3}$. *This completes the discussion of the invariance of Z_{M^3} under Matveev-Piergallini moves.*
  To obtain a topological invariant of M^3, we define

$$I_{M^3,q} = \tau_q^{-t+1} TV_{M^3,q}$$

where t is the number of 3-cells in the decomposition of M^3 corresponding to the given special spine. $I_{M^3,q}$ is the desired invariant of 3-manifolds obtained by state summation.
  It has been shown by Piunikhin [Piu92], by using facts about the quantum group $SL(2)_q$, that $I_{M^3,q}$ coincides precisely with the invariant defined by Turaev and Viro. Our approach provides a direct road to this invariant and to its calculation.

Chapter 11

The Shadow World

11.1 Preliminaries

In this Chapter we consider the link invariants that are obtained by replacing each link component by a parallel cable of strands and placing a
projector in this cable. Thus

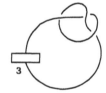

will denote the bracket polynomial of

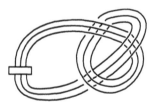

which, in turn, is a sum of bracket polynomials of all the replacements
needed to expand the projector. We shall use the notation $< \vec{a} * K >$ to
denote the bracket polynomial of the link (summation) obtained by placing
an a_i projector on the ith component K_i of a link K. ($\vec{a} = (a_1, \ldots, a_n)$, $K =
K_1 \cup \cdots \cup K_n$).

The evaluations $< \vec{a} * K >$ are invariants of regular isotopy of the link K. Furthermore, we already know, in principle, how to calculate them by using recoupling theory:

(i) Each crossing can be replaced by a sum over trivalent vertex nets via the identity

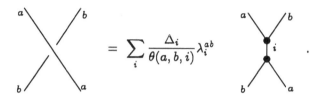

(sums are either generic or over q-admissibles for $q = e^{i\pi/r}$.)

(ii) The basic recoupling formula

can be used repeatedly to reduce the computations to graphs of the form

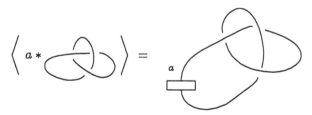

(We leave the second statement as an exercise for the reader.)

Example.

$$\left\langle a * \vcenter{\hbox{\includegraphics{trefoil}}} \right\rangle = \vcenter{\hbox{\includegraphics{shadow}}}$$

$$= \sum_{i,j,k} \frac{\Delta_i \Delta_j \Delta_k}{\theta(a,a,i)\theta(a,a,j)\theta(a,a,k)} \overline{\lambda}_i^{aa} \overline{\lambda}_j^{aa} \overline{\lambda}_k^{aa}$$

$$= \sum_{i,j,k} \frac{\Delta_i \Delta_j \Delta_k}{\theta(a,a,i)\theta(a,a,j)\theta(a,a,k)}$$

$$\overline{\lambda}_i^{aa} \overline{\lambda}_j^{aa} \overline{\lambda}_k^{aa} \frac{\theta(a,a,i)}{\Delta_i} \delta_{ij} \frac{\theta(a,a,i)}{\Delta_i} \delta_{ik} \theta(a,a,i)$$

$$= \sum_{i \in ADM\ (a,a,i)} \Delta_i (\overline{\lambda}_i^{aa})^3.$$

(Here $ADM(a,b,c)$ means either generically admissible or q-admissible if $q = e^{i\pi/r}$.)

In many cases, the $6j$ symbols are not required for the evaluation. They will occur in general, and we seek a general formula for $< \vec{a} * K >$. As we shall see, there is a general formula, and it takes the form of a partition function on a 2-cell complex that is exactly analogous to the partition function that we used on special spines to produce the Turaev-Viro invariant.

11.2   Shadow Translations

In order to obtain the above mentioned partition function the appropriate technique is the translation of recoupling into the "shadow world" of Kirillov and Reshetikhin [KR88]. These shadow ideas of the aforementioned authors have been further developed in a somewhat different direction by Turaev [Tur92]. Here we utilize the Kirillov-Reshetikhin shadow world in the context of our recoupling theory.

What is this shadow world? It is remarkable translation of formalisms where $< \vec{a} * K >$ is transmuted into a recoupling pattern. We start with an example.

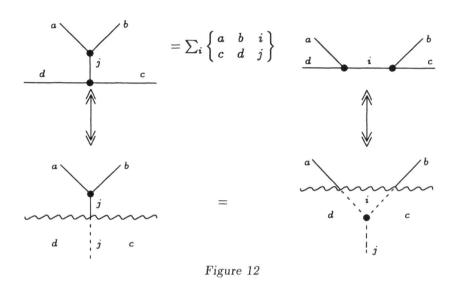

$$= \sum_i \left\{ \begin{array}{ccc} a & b & i \\ c & d & j \end{array} \right\}$$

Figure 12

In the top line of *Figure 12* we have indicated the recoupling formula that defines our $6j$ symbol $\left\{ \begin{array}{ccc} a & b & i \\ c & d & j \end{array} \right\}$. It is drawn so that the legs labelled d and c form a horizontal line, and so that they become (on the right hand side of the formula) the line labelled d-i-c. In the part-shadow-world figures drawn below the recoupling formula the lines d-c and d-i-c have been replaced with an unlabelled "wave-line" of the form ⌇⌇⌇⌇. In the left figure the vertical segment j is extended below the wave-line becoming dotted $(\cdots)$ in the region below the line. The *labels on the horizontal line* d-c *have become region labels below the wave-line. The shadow-world lies below the wave line.*

By moving the wave-line upward we translate a link theoretic formalism into a recoupling formalism. Thus, in the lower right-hand diagram of Figure 12 the wave-line is translated upward so that the 3-valent vertex goes into the shadow world. A new triangular region appears in the shadow world, and this is labelled i. Above the line, we see the recoupling pattern of the right-hand formula. *Below the line, we can decode the pattern of the recoupling 6j coefficients.*

The rule for this decoding is exactly the rule we have used in Chapter 10 for associating recoupling coefficients to labelled 2-cell complexes:

(i)

$$= Tet \begin{bmatrix} a & b & i \\ c & d & j \end{bmatrix}$$

<u>vertex</u>

(ii)

$$Tet \begin{bmatrix} a & b & i \\ c & d & j \end{bmatrix}$$

<u>vertex</u>

(iii)

$$\theta(a, b, c)^{-\chi(e)}$$

<u>edge</u> e

(iv)

$$\Delta_i^{\chi(f)}$$

<u>face</u> f

In the planar pictures, we had previously imagined two-cells abutting to the edges (dotted lines) drawn in the plane. Here (in the shadow world) these edges receive labels, but are not necessarily associated with cells. (Such cells will reappear in the next chapter – associated with a surgery presentation of a 3-manifold.) In any case, trivalent and quadrivalent vertices are associated with tetrahedral symbols as shown in (i) and (ii) above. To an edge e labelled b with adjoining regions labelled a and c is associated $\theta(a, b, c)^{-\chi(e)}$ where $\chi(e) = 0$ if e has no nodes and $\chi(e) = 1$ otherwise. To a face f is associated $\Delta_i^{\chi(f)}$ where $\chi(f)$ is the Euler characteristic of f.

To a planar diagram D in the shadow world, we associate a partition function Z_D consisting in the sum over all q-admissible (or generically admissible) labellings of the products of these weights associated to vertices, edges and faces of the complex. (For knot-theoretic purposes we shall shortly modify the 4-vertex to take care of the two crossing types.)

Now return to Figure 12. The lower right hand diagram

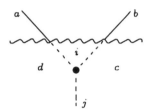

encodes the (partial) partition function

$$\sum_i \frac{Tet \begin{bmatrix} a & b & i \\ c & d & j \end{bmatrix} \Delta_i}{\theta(a,i,d)\theta(b,i,c)} \cdot \Gamma(a,b,c,d,\ldots)$$

(leaving c, d and j fixed) where $\Gamma(a,b,c,d,\ldots)$ denotes the partition function on the rest of the shadow-world part of the graph. Since

$$\begin{Bmatrix} a & b & i \\ c & d & j \end{Bmatrix} = \frac{Tet \begin{bmatrix} a & b & i \\ c & d & j \end{bmatrix} \Delta_i}{\theta(a,i,d)\theta(b,i,c)},$$

we see that we can regard the symbolic identity

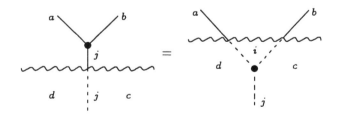

as an actual identity of partition functions *and* that it is equivalent to the recoupling formula at the top of Figure 12.

11.3   Proving Shadow World Transitions

We now proceed to give the translations for a sufficient number of graphic situations to allow for a complete transition of a link diagram into the shadow world. In order to do this task, it is sometimes useful to adopt a hybrid formalism that is already implicit translation to the shadow world. In the hybrid formalism we regard the wave-line as both a recoupling *and* as the boundary line between the worlds of light (above) and shadow (below). Thus, in

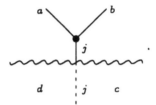

The wave-line is interpeted as *both* a division between light and shadow *and* as a pair of recoupling lines labelled d and c. The wave-line acquires its labels from the labels of the shadow regions below it.

We can then apply the recoupling formula and write:

$$= \sum_i \qquad \qquad \left\{ \begin{matrix} a & b & i \\ c & d & j \end{matrix} \right\}$$

$$=$$

The second inequality absorbs the summation by our conventions about the partition function associated with the shadow part of the diagram. Note that the labels on the wave-line are finally induced entirely from below. With this formalism in hand, we can state and prove the desired identities.

$1^{\underline{0}}.$

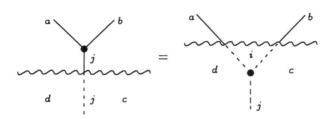

Proof: (given above). ∎

$2^{\underline{0}}$

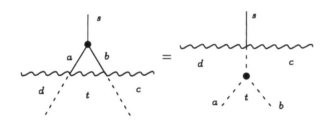

Proof:

$$\text{[triangle diagram]} = \sum_i \left\{ \begin{matrix} d & a & i \\ b & c & t \end{matrix} \right\} \text{[loop diagram]}$$

$$= \sum_i \left\{ \begin{matrix} d & a & i \\ b & c & t \end{matrix} \right\} \frac{\theta(a,b,s)}{\Delta_s} \delta_{is} \text{[vertex diagram]}.$$

Thus

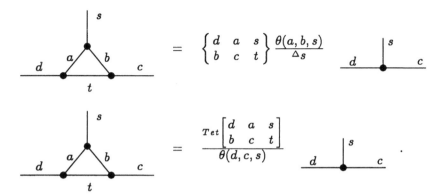

Translated into hybrid formalism, this recoupling identity becomes the assertion we wanted to prove.   ∎

$3^{\underline{o}}$

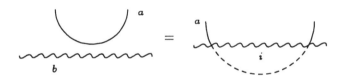

Proof:

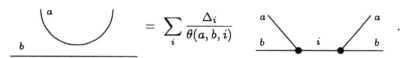

This recoupling formula translates as indicated.   ∎

$4^{\underline{o}}$

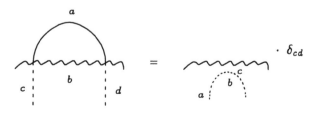

Proof:

$$c \overset{a}{\underset{b}{\frown}} d = \frac{\theta(a,b,c)}{\Delta_c}\delta_{cd} \qquad c \qquad .$$

Note that in this instance the correct assignments of weights follows the rules $\Delta^{\chi(face)}$, $\theta^{-\chi(edge)}$. ∎

The next identity shows how to pass a link theoretic crossing into the shadow world. In order to include crossings we need to indicate two types of vertex in the shadow world. We consider a bicoloration of the faces of the projections of the links, with the outer face being white. A *backslash crossing* is a crossing in which going from one black face to its opposite through the vertex the overpass is from northwest to southeast. The other type is a *slash crossing*. To avoid the bicoloration of the shadow world diagrams, before diving into the shadow world, we effect a rotation, if necessary, so that the black faces are the ones in the top and in the bottom while the white faces are in the left and right. So we write these nodes in the forms:

(slash and back-slash nodes).

We assign an extra vertex weight to these nodes as follows:

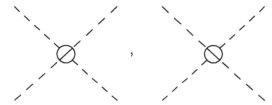

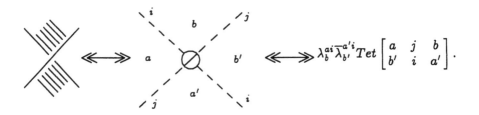

Here $\lambda_c^{ab} = (-1)^{(a+b-c)/2} A^{[a(a+2)+b(b+2)-c(c+2)]/2}$ as in Section 9.9, and all the triples (a, i, b), (a', i, b') are q-admissible (or admissible) following our prior conventions. It is also understood that these crossings are oriented with respect to horizontal and vertical (by the convention about the face coloration) relative to the wave-line above them. $\overline{\lambda}$ denotes the multiplicative inverse of λ.

5°

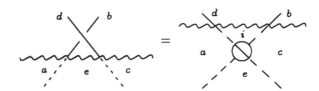

Proof:

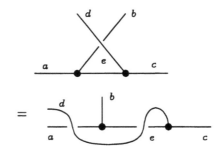

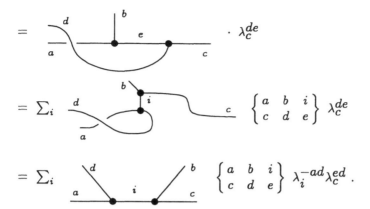

$$= \quad \cdot \; \lambda_c^{de}$$

$$= \sum_i \quad \begin{Bmatrix} a & b & i \\ c & d & e \end{Bmatrix} \lambda_c^{de}$$

$$= \sum_i \quad \begin{Bmatrix} a & b & i \\ c & d & e \end{Bmatrix} \lambda_i^{-ad} \lambda_c^{ed} \; .$$

By the conventions just prior to $5^{\underline{o}}$, this completes the proof. ∎

With these five translation formula, we can now re-write the calculation of $< \vec{a} * K >$ as a shadow world partition function. Since $\Delta_0 = 1$ we have

$$D = \underset{0}{\overbrace{\hspace{3cm}}^{\,D\,}} \; = \; \overset{\frown}{\underset{0 \qquad D'}{}} \quad ,$$

where D' is a dotted form of D, as the general form of the evaluation.

11.4 Examples

Example 1. If $b = 0$, then

The left-hand side is $\Delta_a \Delta_b$. The right-hand side is

$$\sum_{i \in ADM(a,b,i)} \Delta_i.$$

That $\Delta_a \Delta_b = \sum_{i \in ADM(a,b,i)}$ follows from the recoupling:

$$= \sum_i = \sum_{i \in ADM(a,b,i)} \frac{\Delta_i}{\theta(a,b,i)}$$

$$\Rightarrow \bigcirc a \quad \bigcirc b \; = \; \sum_{i \in ADM(a,b,i)} [\Delta_i / \theta(a,b,i)] \quad a \bigcirc i \; b$$

$$\Rightarrow \Delta_a \Delta_b = \sum_{i \in ADM(a,b,i)} \Delta_i .$$

If $b = 0$, then both sides are Δ_a.

Example 2.

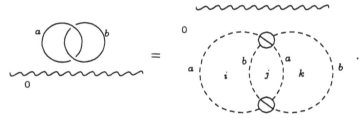

This implies the formula

$$a \bigotimes b \; = \; \sum_{\substack{i, j, k \\ ADM(0,a,i) \\ ADM(i,b,j) \\ ADM(j,a,k) \\ ADM(k,b,0)}} \frac{\Delta_i \Delta_j \Delta_k}{\theta(0,a,i)\theta(i,b,j)\theta(j,a,k)\theta(k,b,0)} \overline{\lambda}_0^{ia} \lambda_k^{ja} \, \overline{\lambda}_j^{ib} \lambda_k^{0b}$$

$$TET \begin{bmatrix} i & b & 0 \\ k & a & j \end{bmatrix} TET \begin{bmatrix} i & a & j \\ k & b & 0 \end{bmatrix} .$$

Thus $b = k, i = a$ and we get $\theta(0,a,i) = \Delta_a$, $\theta(k,b,0) = \Delta_b$, $TET \begin{bmatrix} i & b & 0 \\ k & a & j \end{bmatrix} = \theta(a,b,j) TET \begin{bmatrix} i & a & j \\ k & b & 0 \end{bmatrix} = \theta(a,b,j)$

$$\overline{\lambda}^{ia} = \overline{\lambda}^{aa}$$

$$\lambda_k^{ja} = \lambda_b^{ja}$$

$$\overline{\lambda}_j^{ib} = \overline{\lambda}^{ab}$$

$$\lambda_k^{0b} = \lambda_b^{0b} = 1.$$

Thus

$a \quad b$ $\quad = \quad \sum_{j \in ADM(a,b,j)} \frac{\Delta_j}{\theta(a,b,j)^2} \overline{\lambda}^{aa} \lambda_b^{ja} \overline{\lambda}_j^{ab} \theta(a,b,j)^2$

$$= \quad \sum_{j \in ADM(a,b,j)} \Delta_j \lambda_b^{ja} \overline{\lambda}_j^{aa} \overline{\lambda}_0^{aa}$$

Letting $x' = x(x+2)$, we have

$$\lambda_c^{ab} = (-1)^{\frac{a+b-c}{2}} A^{\frac{a'+b'-c'}{2}}$$

thus

$$\lambda_b^{ja} \overline{\lambda}_j^{ab} \overline{\lambda}_0^{aa} = (-1)^{\frac{i+a-b}{2}} (-1)^{\frac{a+b-j}{2}} (-1)^a A^{\frac{i'+a'-b'}{2}} A^{\frac{i'-a'-b'}{2}}$$

$$A^{-a'} = A^{j'-a'-b'}.$$

Thus we conclude that

$a \quad b$ $\quad = \quad \sum_{j \in ADM(a,b,j)} \Delta_j A^{j(j+2)-a(a+2)-b(b+2)}$.

The reader may enjoy comparing this formula with the direct formula that is easily derived from Chapter 9.

We can summarize our discussion by the

Theorem 5 *The $\vec{a} * K$ denote a link diagram equipped with an a_i-projector parallel cable on each component K_i. Let $D(K, \vec{a})$ denote the diagram of K with label a_i on component K_i. Arrange $D(K, \vec{a})$ with respect to the vertical so that the only critical points are local maxima, minima and crossings. Let $ShD(K, \vec{a})$ denote the shadow diagram obtained from $D(K, \vec{a})$ by sweeping a wave-line upward from below $D(K, \vec{a})$. Let $Z(K, \vec{a}; b)$ denote the partition function for $ShD(K, \vec{a})$ where the outer region has fixed label b, all edges* *acquire fixed labels from* $(K, \vec{a})$ *and the regions have variable labels* $\vec{c}$. *(We say that* $\vec{\imath}$ *is admissible if all components of* $\vec{\imath}$ *satisfy the local admissibility conditions specified by their locations in the shadow.) Vertex weights for* $Z(K, \vec{a}; b)$ *are as described in this chapter. Then*

$$< \vec{a} * K >= \frac{1}{\Delta_b} Z(K, \vec{a}; b).$$

Discussion. This theorem gives a specific formula for the evaluation of "heirarchical Jones polynomials". (See [KM92]). The reader may wonder what would happen if we also summed over evaluations for different $\vec{a}$. The partition function Z would then have a form similar to the one for the Turaev-Viro invariant. At roots of unity, we shall do precisely this in the next Chapter. The result is a formula for the $SU(2)$ 3-manifold invariant of Witten-Reshetikhin-Turaev.

Chapter 12

The Witten-Reshetikhin-Turaev Invariant

In this Chapter we show how the recoupling theory gives the existence of an invariant of oriented 3-manifolds via surgery on links [Lic62] and Kirby calculus [Kir78]. This is the Witten-Reshetikhin-Turaev Invariant [RT91], [Wit89] for the group $SU(2)$. We use the approach via Temperley-Lieb algebra as pioneered by Lickorish [Lic91]. By combining the Temperley-Lieb algebra approach with the recoupling theory one obtains suprisingly simple proofs of existence of the invariant plus new reformulations of it via the shadow translations of Chapter 11. Our treatment of this shadow translation of the Witten-Reshetikhin-Turaev Invariant owes a debt to the paper [Tur92] by Turaev, but here the translation is done very directly in terms of our elementary reformulation of the recoupling theory. As a preliminary to constructing the invariant, we quickly recall how to construct 3-manifolds by surgery on framed links.

12.1 Framed Links

Framed links are represented by regular isotopy classes of link diagrams modulo *ribbon equivalence* where ribbon equivalence is generated by the relation $\searrow\hspace{-0.3em}\bigcirc \approx \bigcirc\hspace{-0.6em}\diagup$. With this extra relation, the ribbon equivalence classes of link diagrams represent framed links via the *blackboard framing*. The blackboard framing is obtained by placing a vector field normal to each link component with all the normals in the plane indicated by the diagram. We can indicate this framing by sketching the tips of the normals as a second component parallel to each component, so that each component is replaced by an immersed band:

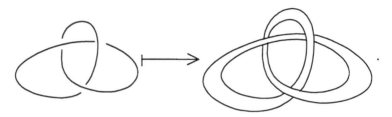

The reason for ribbon equivalence is manifest in that the framings corresponding to these two curls (with different winding number in the plane) are ambient isotopic.

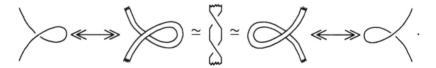

Corresponding to each framed link K, there is a 3-manifold $M^3(K)$ and 4-manifold $W^4(K)$ with $\partial W^4(K) = M^3(K)$ where ∂X denotes the boundary of X. $W^4(K)$ is obtained by attaching 2-handles to the 4-ball D^4 along the components of the link $K \subset \partial D^4 = S^3$. ($S^3$ denotes the three dimensional sphere.) More precisely, a 2-handle is a copy of $D^2 \times D^2$, the cartesian product of two-dimensional disks. To attach a 2-handle, one must be given an embedding $f : S^1 \times D^2 \longrightarrow S^3$. Then since the boundary of $D^2 \times D^2$ is $(S^1 \times D^2) \cup (D^2 \times S^1)$ one can form the identification space $D^4 \cup_f (D^2 \times D^2)$ where f attaches the $S^1 \times D^2$ part of the boundary of $D^2 \times D^2$ to $S^3 = \partial D^4$. The space $D^4 \cup_f (D^2 \times D^2)$ is said to be the result of attaching a handle to D^4 via f.

Each component of a framed link specifies a mapping $f : S^1 \times D^2 \longrightarrow S^3$. The framing determines the extension from $S^1 \times 0 \longrightarrow S^3$ to $S^1 \times D^2 \longrightarrow S^3$. In the case of the blackboard framing, this means that the linking number of $f_K(S^1 \times 1)$ with $f_K(S^1 \times 0)$ [regard D^2 as the unit disk in the complex plane] is equal to the writhe, $w(K)$, for the given component K. Thus, in the blackboard framing, the writhe $w(K)$ is the framing number of the component K.

By attaching handles to each component K of a framed link L we obtain a 4-manifold $W^4(L)$ consisting in D^4 with one $D^2 \times D^2$ attached for each K in L. The boundary, $M^3(L) = \partial W^4(L)$ is said to be the 3-manifold obtained by doing surgery on the framed link L. $M^3(L)$ can be described directly by the prescription: Cut out a copy of $S^1 \times D^2$ for each component K of L and sew in a copy of $D^2 \times S^1$ so that the meridian $m = \{(\lambda, 1)|\ |\lambda| = 1\} \subset D^2 \times S^1$ is attached to $f_K(S^1 \times 1)$, the *longitude* of the framed link component K. Note that this means that this longitude $l_K = f_K(S^1 \times 1)$ will bound a disk in $M^3(L)$.

12.2   Examples

Example 1. Suppose that L consists of a single component with framing number 3.

L

Then $M^3(L)$ is obtained by sewing $m \subset D^2 \times S^1$ to $l_K \subset S^3 - (S^1 \times D^2)^0$ ($X^0 = $ interior of X) as shown below.

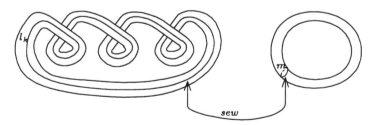

sew

Since l_K represents thrice the generator of $H_1(S^3 - (S^1 \times D^2)^0)$, it is easy to see that $H_1(M^3(L)) \cong \mathbb{Z}/3\mathbb{Z}$. In fact, $M^3(L)$ is a lens space of type $(3, 1)$, in this example.   ∎

More generally, it is easy to read a presentation of $G(L) = \pi_1(M^3(L))$, the fundamental group of the 3-manifold $M^3(L)$. The result is that $G(L) = \pi_1(S^3 - L)/ < l(L) >$ where $< l(L) >$ is the normal subgroup of the fundamental group of $S^3 - L$ generated by the longitudes $[l_K]$ of the components K of L (referred to a suitable basepoint). In the blackboard framing, we can specify the elements $[l_K]$ in the Wirtinger presentation as described below.

In the Wirtinger presentation, one chooses an orientation for the link diagram, and associates one generator for each arc in the diagram and one relation at each crossing of the forms $c = b^{-1}ab$ or $c = bab^{-1}$ according as the crossing is right or left handed, respectively. Here b denotes the over-crossing arc, a is the incoming under cross and c is the out going under cross.

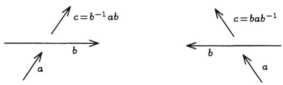

The fundamental group $\pi_1(S^3 - L)$ is given by these generators and relations.

In these terms, the longitude $\lambda_K = [l_K] \in \pi_1(S^3 - L)$ is given by choosing a point on K, walking along K in the direction of its orientation and writing the product of the elements of $\pi_1(S^3 - L)$ encountered as over-crossing arcs during a once-round K journey. This product is λ_K, and it is well-defined up to conjugacy.

Example 2.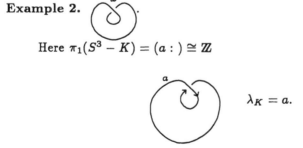

Here $\pi_1(S^3 - K) = (a :) \cong \mathbb{Z}$

$\lambda_K = a.$

Hence $G(K) = \pi_1(M^3(K)) = \{1\}$. In fact, it is easy to see that $M^3(K)$ is homeomorphic to S^3. This is the fundamental case of a surgery whose result does not change the topological type of the corresponding 3-manifold.

Example 3.  K

$$\pi_1(S^3 - K) = (a:), \quad \lambda_K = 1$$
$$G(K) = (a:).$$

Here $M^3(K) \cong S^2 \times S^1$.

Example 4.

$$
\begin{aligned}
c &= b^{-1}ab \\
b &= a^{-1}ca \\
a &= c^{-1}bc \\
\lambda_K &= cbac^{-2}
\end{aligned}
$$

Thus $G(K) = (a, b, c : c = b^{-1}ab, \ b = a^{-1}ba, \ a = c^{-1}bc, \ cbac^{-2} = 1)$. Since $H_1(M^3(K))$ is the abelianization of $G(K)$ we find (writing the relations for H_1 additively): In $H_1(M^3(K))$

$$
\begin{aligned}
c &= a \\
b &= c \\
a &= b \\
0 &= a + b - c = a.
\end{aligned}
$$

Hence $H_1(M^3(K))$ is the trivial group. However, $G(K)$ is not trivial. It is isomorphic with the binary dodecahedral group in $SU(2)$ and $M^3(K)$ is the dodecahedral space, obtained by gluing opposite faces of a dodecahedron by $2\pi/5$ twists. ∎

With those examples in mind, we can state the well-known

Poincaré Conjecture. If $\pi_1(M^3) = \{1\}$ then M^3 is homeomorphic to the 3-sphere S^3. (M^3 a compact orientable 3-manifold).

It is to be hoped that the 3-manifold invariants studied herein will eventually shed light on this conjecture.

12.3   Handle Sliding and Kirby Calculus

We now turn to the way homeomorphism classes of 3-manifolds are represented by links in the blackboard framing. There is a move on link diagrams that corresponds to the sliding of handles in the 4-manifold $W^4(L)$. For links the move is as illustrated below:

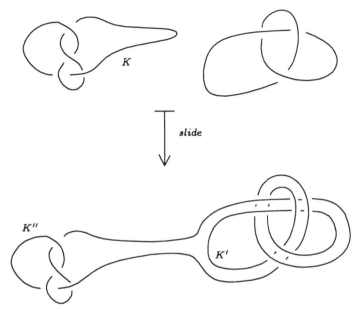

In a handle-slide (at link level) one component (K) is replaced by a new component (K'') obtained by taking a connected sum of K with a parallel push-off (in the blackboard framing) of another component K'. The operation is reversible. The connected sum is obtained by moving the diagrams by regular isotopy so that an arc of K is contignous with an arc of K' as shown below.

Thus, we can indicate a handle-slide symbolically via the heiroglyph:

We then have the following basic results:

Theorem 6 *[Lickorish]. Every compact orientable 3-manifold can be represented as surgery on a framed link $L \subset S^3$.*

Theorem 7 *[Kirby]. $M^3(L)$ is homeomorphic to $M^3(L')$ if and only if L' can be obtained from L by combinations of*

1) *regular isotopy and ribbon equivalence.*

2) *handle slides*

3) *addition and subtraction (blowing up and blowing down) of disjoint*

components of the form  .

Here L is a link diagram in blackboard framing.

Remark 15 *Kirby did not use the blackboard framing. The approach using it is very helpful for our purposes.*

The Kirby and Lickorish Theorems reduce the study of 3-manifolds to a subtle chapter in the theory of knots and links. There are a number of consequences of elementary handle slides that are of use. We illustrate a few now.

12.4 Consequences of Handle Slides

$1^{\underline{o}}$

in general, if $\Big|\Big| \cdots \Big|$ denotes a parallel cable of n strands, then

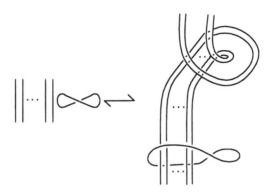

Since $\Big|\Big| \cdots \Big| \, \rightharpoonup \, \Big|\Big| \cdots \Big| \, \infty$, we can write

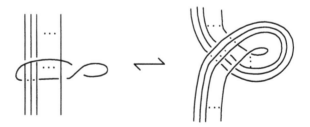

A single unknotted curve of framing ± 1 encircling a cable of parallel lines can be replaced by a $\mp$ twist of 2π in the cable.

$2^{\underline{o}}$

Crossings can be switched at the expense of adding more unknotted components. Hence every 3-manifold is represented by a link whose components are unknotted.

$3^{\underline{o}}$ A component C of a link L is a *circumcised component* if there is another component D unknotted, separated from any other component of L by an S^2 and having linking number ± 1 with C. D is called the *encircling companion* of C.

Any crossing in a circumcised component of a blackboard framed link can be switched at the cost of adding a curl in the component. A *circumcision pair* is a circumcised component together with its encircling companion.

Corollary 3 *Dropping a circumcision pair P from a blackboard framed link L does not change* $M^3(B)$.

Proof: By switching crossings (and adding curls) we can free and unlink the circumcision pair; that is, we arrive to an equivalent link obtained from L by dropping P and repacing it by the pair

separated by an S^2 from the rest of the link.

Note that curls in a circumcised component can be switched:

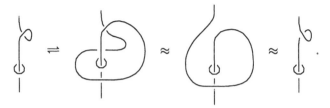

Therefore, we get an equivalent link with one negative or no curls:

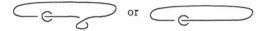

However, $S^3 \cong S^3 \# S^3$ and

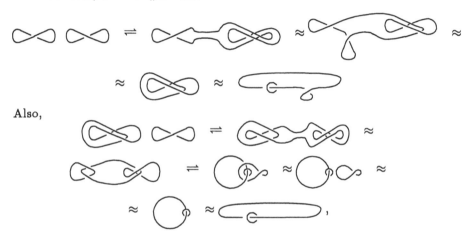

Also,

proving the Corollary. ∎

Corollary 4 *If a framed link L is made of n circumcised components together with their matching n encircling companions, then $M^3(L) \cong S^3$.*

Proof: Repeat n times the previous Corollary ∎

This Corollary implies that any 3-manifold induced by an n-component link can be obtained from a link inducing $\#_n(S^2 \times S^1)$ by switching n crossings.

A combination of observations now leads to the theorem of Fenn and Rourke – that the moves of type $1^{\underline{o}}$ plus blowing up and down can generate all handle-sliding. See [FR79] and also [Kau91] for a diagrammatic proof.

It is now easy to create many strange presentations of S^3:

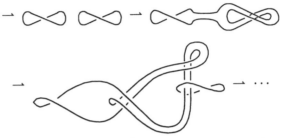

We know that all presentations of S^3 are obtained by blowing up and down and handle sliding from nothing. The Poincaré Conjecture states that if $G(L) = \{1\}$, then L can be obtained from nothing in this way! Is it true??

12.5   Invariants

We are now in a position to discuss the construction of invariants of 3-manifolds via surgery and the Kirby calculus. Our first construction works directly by recoupling theory. This approach is based on ideas of Justin Roberts and Oleg Viro. The specific way we phrase the argument is our own formulation.

We shall consider an invariant $<< K >>$ of *labelled* link diagrams with trivalent vertices. We shall assume that the labels $a, b, c, \ldots$ belong to a specific finite index set $\mathcal{I}$ and that they label entire link components or, in the presence of vertices, arcs (including under crossings) from vertex to vertex. We say that $<< K >>$ is an invariant of *ribbon equivalence* if $<< K >>$ is an invariant of

1) regular isotopy.

2) ribbon equivalence in the sense

$$\left\langle\!\!\left\langle \; \overset{a}{\curvearrowright}\!\bigcirc \; \right\rangle\!\!\right\rangle = \left\langle\!\!\left\langle \; \bigcirc\!\diagup^a \; \right\rangle\!\!\right\rangle$$

for any label a.

3)

for any labels a, b, c, d.

(It is possible that there are admissibility criteria for labelling at vertices. It is assumed that $<< K >>$ vanishes at inadmissible labellings). Call $<< K >>$ a *labelled ribbon graph invariant.*

Example Let $\vec{a}$ denote the labelling on K and let $<< K >>=< \vec{a} * K >$ where $A = e^{i\pi/2r}$, $\mathcal{I} = \{0, 1, 2, \ldots, r - 2\}$ and the conventions for trivalent vertices and admissibility are as constructed in this monograph. This is our basic collection of examples of ribbon invariants.

Lemma 21 *Suppose that the ribbon graph invariant* $<< K >>$ *satisfies the identity*

where $\varphi(a, b, c)$ *is a function on* $\mathcal{I}^3$, *symmetric in* (a, b, c), $\{c_i\}$ *are constants, and a and b are arbitrary labels. Let* $\Big)^{\omega}$ *denote the sum*

$$\Big)^{\omega} = \Big(\sum_{i \in \mathcal{I}} c_i \Big)^{i}$$

and let $\ll w * K \gg$ denote the sum of (invariants of) labellings, weighted by products of c_i's that is obtained by placing an w on each component of K.

$$\left[e.g. \quad {}^{\omega}\bigcirc\hspace{-0.3em}\bigcirc{}^{\omega} \quad = \sum_{i,j \in \mathcal{I}} c_i c_j \quad {}^{i}\bigcirc\hspace{-0.3em}\bigcirc{}^{j} \right]$$

Then $\ll w * K \gg$ is invariant under handle-sliding.

Remark 16 *Let $< w * K >$ denote $\ll w * K \gg$ in the case of our basic examples with $A = e^{i\pi/2r}$, $\mathcal{I} = \{0, 1, 2, \ldots, r-2\}$. The recoupling theory tells us (see Section 9) that*

$$\left.{}^{a}\right)\hspace{1em}\left({}^{b}\right. = \sum_{i \in \mathcal{I}} \frac{\Delta_i}{\theta(a,b,i)} \quad \overset{a\quad b}{\underset{a\quad b}{\times_i}}$$

where Δ_i and $\theta(a, b, i)$ are defined in Chapter 3 and Chapter 5. Note that by Corollary 2, $\theta(a, b, i) \neq 0$ when the sum is over q-admissible labellings.

Note also that $\left[\left.\right)^{a} \right]$ *will always indicate the presence of a projector:*

$$\left.\right)^{a} = \overset{a}{\boxed{}} \quad \text{in this context.}$$

Thus we have that $< w * K >$ is an invariant of handle sliding by this lemma.

Proof of Lemma We shall illustrate the proof (brackets $\ll\gg$ deleted, but understood) by handle-sliding over a trefoil diagram. The argument implicit in these pictures is completely general.

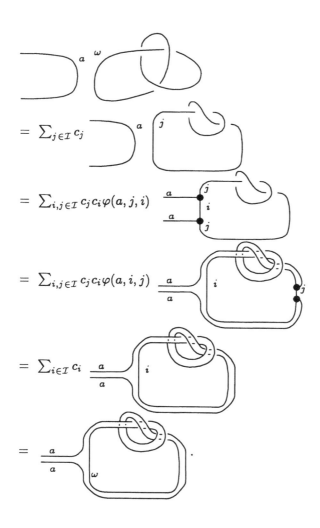

This completes the proof. ∎

12.6   Lickorish's Proof

Remark 17 *Along with the recoupling formula, the basic ingredient of the proof of Lemma 21 is the isotopy of embedded graphs:*

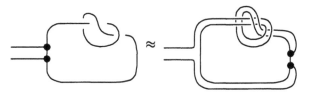

The proof suggests that we do not need the full force of the recoupling theory, and that there may be other significant examples that fall under the aegis of this formalism. The latter point is left to futher investigation. However, the former point is indeed true as exemplified by the following startlingly simple proof of invariance of $< w*K >$ under handle slides due to Lickorish [Lic93].

Observe that the recursion formula

$$\Delta_{n+1} \begin{array}{c} n+1 \mid \quad \mid 1 \\ \boxed{} \\ \mid \quad \mid \end{array} = \Delta_{n+1} \begin{array}{c} n+1 \mid \quad \mid 1 \\ \boxed{} \\ \mid \quad \mid \end{array} + \Delta_n \begin{array}{c} n \quad \mid 1 \quad \mid 1 \\ \boxed{} \\ \boxed{} \\ \mid 1 \quad \mid 1 \end{array}$$

implies (for $A = e^{i\pi/2r}$, and note that $\Delta_{r-1} = 0$, $\begin{array}{c} r-1 \\ \boxed{} \\ \mid \end{array} = 0$) that

$$\sum_{n=0}^{r-2} \Delta_n \begin{array}{c} n \quad \mid 1 \\ \boxed{} \\ \mid \end{array} = \sum_{n=0}^{r-2} \Delta_n \begin{array}{c} n \mid \quad \mid 1 \\ \boxed{} \\ \mid \quad \mid \end{array} + \sum_{n=0}^{r-2} \Delta_n \begin{array}{c} n \quad \mid 1 \quad \mid 1 \\ \boxed{} \\ \boxed{} \\ \mid 1 \quad \mid 1 \end{array} .$$

We can abbreviate this formula as:

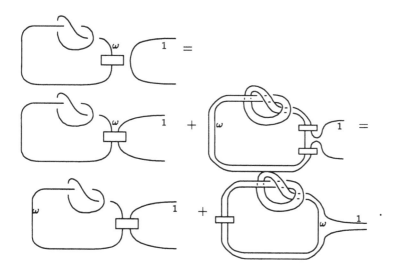

where it is understood that

Then

Hence, by symmetry,

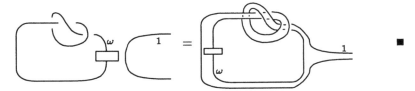

  This proof illuminates the fundamental role of the Jones-Wenzl recursion relations (for projectors) in this theory.

12.7   Normalization

We now let $A = e^{i\pi/2r}$ and consider the invariant of handle-slides, $< w * K >$. In order to normalize this to an invariant of the 3-manifold $M^3(K)$, choose an orientation for the link K and let $N = N(K)$ be the matrix given by the entries:

$$N_{ij} = \begin{cases} \text{link } (K_i, K_j) & \text{for } i \neq j \\ wr(K_i) & \text{for } i = j \end{cases}$$

where K has components $K_1, K_2, \ldots, K_n$ and link (x, y) denotes the linking number of x with y. ($N(K)$ is the intersection form on $H_2(W^4(K))$ with respect to an appropriate basis choice.). Let $b_+(K)$ denote the number of positive eigenvalues of $N(K)$ and $b_-(K)$ the number of negative eigenvalues.

Finally, let $n(K) = b_+(K) - b_-(K)$. Note that $n \left(\vcenter{\hbox{$\omega$}} \right) = +1$ and

$n \left(\vcenter{\hbox{$\omega$}} \right) = -1$. Therefore, *define*

$$Z'(M^3(K)) = < w * K > \left\langle \vcenter{\hbox{ω}} \right\rangle^{-b_+(K)} \left\langle \vcenter{\hbox{ω}} \right\rangle^{-b_-(K)} .$$

It follows that $Z'(M^3)$ is an invariant of the 3-manifold $M^3(K)$ and it is normalized so that $Z'(S^3) = 1$.

For purposes of working with topological quantum field theory ([FG91], [Wit89]) it makes sense to normalize in a way so that $S^2 \times S^1$ has invariant 1. Let $\Lambda_1 = \left\langle \vcenter{\hbox{}} \right\rangle$. (We shall calculate this shortly.) Let $\mu = \sqrt{\frac{2}{r}} sin(\pi/r)$. Let $|K|$ denote the number of components of K. *Define* $Z_K = Z(M^3(K)) = < w * K > \mu^{|K|+1} \alpha^{-n(K)}$ where $\alpha = \mu\Lambda_1$. We shall see, as we do some calculations, that Z_K is the normalization that gives $Z(S^2 \times S^1) = 1$ and $Z(S^3) = \mu = \sqrt{\frac{2}{r}} sin(\pi/r)$.

Calculations The following lemma due to Lickorish [Lic93] is useful.

Lemma 22 *(Encirclement Lemma). Let* $A = e^{i\pi/2r}$. *Then, for* $n \neq 0$,

$= 0$. *That is,*

$= 0$ *for any n-tangle T.*

Proof: $= -(A^{2(n+1)} + A^{-2(n+1)})$ (Prop. 7).

Therefore,

$$(-A^2 - A^{-2}) \quad \ \quad = \quad \ \quad = \quad \$$

$$= \quad -(A^{2(n+1)} + A^{-2(n+1)}) \quad \ \quad .$$

This proves the lemma. ∎

Example $\left\langle \ \right\rangle$

$$= \ = \$$

$$= \ = \ \quad .$$

Hence

$$\left\langle \ \right\rangle = \left\langle \ \right\rangle$$

$$= \left\langle \ \right\rangle$$

$$= \sum_{i=0}^{r-2} \Delta_i^2 = -2r/(A^2 - A^{-2})^2.$$

with $A^2 = e^{i\pi/r}$, $A^2 - A^{-2} = 2i\, sin(\pi/r)$. Thus, for $\mu = \sqrt{\frac{2}{r}}\, sin(\pi/r)$ we

have $\left\langle \vcenter{\hbox{}} \right\rangle \left\langle \vcenter{\hbox{}} \right\rangle = \mu^{-2}$. Hence

$$Z(S^3) = Z\left(M\left(\vcenter{\hbox{}} \right) \right) = \left\langle \vcenter{\hbox{}} \right\rangle \mu^2 \alpha^{-1}$$

$$Z(S^3) = \mu.$$

To complete the normalization, we want to find $\alpha = \mu \left\langle \vcenter{\hbox{}} \right\rangle$ explic-

itly. This will then complete the normalization for the invariant evaluated at $A = e^{i\pi/2r}$. In [Lic90] it is noted that by replacing A by $-A$ one gets the invariant with normalizations as defined by Kirby and Melvin [KM91]. From our point of view it is pleasing to be able to use $A = e^{i\pi/2r}$ directly. The evaluation, however, depends upon an extra subtlety with Gauss sums. The result is

Lemma 23 *For $A = e^{i\pi/2r}$,* $\left\langle \vcenter{\hbox{}} \right\rangle = \mu^{-1}(-i)^{r-2}e^{i\pi\left[\frac{3(r-2)}{4r}\right]}$.

Hence, $\alpha = (-i)^{r-2}e^{i\pi\left[\frac{3(r-2)}{4r}\right]}$.

Proof: Let $\Lambda_1 = \left\langle \vcenter{\hbox{}} \right\rangle$. Then

$$\Lambda_1 = \sum_{i=0}^{r-2} \Delta_i \left\langle \vcenter{\hbox{}} \right\rangle = \sum_{i=0}^{r-2} \Delta_i^2 (-1)^i A^{i(i+2)}$$

$$= \sum_{a=1}^{r-1} \Delta_{a-1}^2 (-1)^{a-1} A^{a^2-1}$$

$$\Lambda_1 = -A^{-3} \sum_{a=1}^{r-1} \Delta_{a-1}^2 (-A)^{a^2+2}.$$

Let $B = -A$, And $x = -A^3\Lambda_1$. Then

$$
\begin{aligned}
x &= \sum_{a=1}^{r-1} \Delta_{a-1}^2 B^{a^2+2} = \tfrac{1}{4}\sum_{a=1}^{4r} \Delta_{a-1}^2 B^{a^2+2} \\
&= \frac{1}{4(A^2 - A^{-2})^2}\sum_{a=1}^{4r}(B^{2a} - B^{-2a})^2 B^{a^2+2} \\
&= \frac{1}{4(A^2 - A^{-2})^2}\sum_{a=1}^{4r} B^{-2}(B^{(a+2)^2}) + B^{-2}(B^{(a-2)^2}) - 2B^2(B^{a^2}) \\
&= \frac{2(B^{-2} - B^2)}{4(A^2 - A^{-2})^2}\sum_{a=1}^{4r} B^{a^2} \\
&= \frac{-1}{2(A^2 - A^{-2})}\sum_{a=1}^{4r} B^{a^2}.
\end{aligned}
$$

Claim $\sum_{a=1}^{4r} B^{a^2} = (-i)^r 2\sqrt{2r}e^{i\pi/4}$ when $B = -e^{i\pi/2r}$. We defer the proof of this claim momentarily.

Thus $x = \dfrac{-1}{2(A^2 - A^{-2})}(-i)^r 2\sqrt{2r}e^{i\pi/4}$

$$
\begin{aligned}
\Lambda_1 &= -A^{-3}x = \frac{A^{-3}(-i)^r\sqrt{2r}e^{i\pi/4}}{2\,i\,sin(\pi/r)} \\
&= \mu^{-1}(-i)^{r-2}(i)e^{i\pi/4}e^{-3i\pi/4} \\
&= \mu^{-1}(-i)^{r-2}e^{i\pi\left[\frac{3(r-2)}{4r}\right]}.
\end{aligned}
$$

This completes the proof of the Lemma, modulo the *claim*. ∎

12.8   Gauss Sums

Proof of Claim We use the following result about Gauss sums (see [Nag64], pp. 177–180):

$$
\left\{
\begin{aligned}
&\text{If } \varphi(m, n) = \sum_{s=1}^{n} e^{2\pi i\left(\frac{m}{n}\right)s^2} \\
&\text{then (i) } \varphi(1, n) =
\begin{cases}
(1+i)\sqrt{n} & n \equiv 0(\mathrm{mod}\ 4) \\
\sqrt{n} & n \equiv 1(\mathrm{mod}\ 4) \\
0 & n \equiv 2(\mathrm{mod}\ 4) \\
i\sqrt{n} & n \equiv 3(\mathrm{mod}\ 4)
\end{cases} \\
&\quad \text{(ii)}\,\varphi(hm, n)\varphi(hn, m) = \varphi(h, mn) \text{ for } m, n \text{ relatively prime and} \\
&h \text{ any integer.}
\end{aligned}
\right.
$$

With $A = e^{2\pi i/4r}$, $-A = e^{2\pi i \left[\frac{2r+1}{4r}\right]}$. Thus, we wish to show that $\varphi(2r + 1, 4r) = (-i)^r 2\sqrt{2r}e^{i\pi/4}$. Note that $e^{2\pi i(4r/(2r+1))} = e^{2\pi i((2r-1)/(2r+1))}$.
Hence

$$\varphi(2r - 1, 2r + 1) = \varphi(4r, 2r + 1).$$

By (ii) $\varphi(2r + 1, 4r)\varphi(4r, 2r + 1) = \varphi(1, 4r(2r + 1))$. By (i) $\varphi(1, 4r(2r + 1)) = 2(1 + i)\sqrt{r(2r + 1)}$. Thus, letting $\Phi = \varphi(2r + 1, 4r)$, we have (∗) $\Phi \varphi(2r - 1, 2r + 1)2(1 + i)\sqrt{r(2r + 1)}$.

Again by (ii), $\varphi(2r - 1, 2r + 1)\varphi(2r + 1, 2r - 1) = \varphi(1, (2r + 1)(2r - 1))$ but $\varphi(2r + 1, 2r - 1) = \varphi(2, 2r - 1)$.
Hence

$$\Phi = 2(1 + i)\sqrt{r(2r + 1)}\varphi(2, 2r - 1)/\varphi(1, (2r + 1)(2r - 1)).$$

By (i) $\varphi(1, (2r + 1)(2r - 1)) = i\sqrt{(2r + 1)(2r - 1)}$. It remains to compute $\varphi(2, 2r - 1)$.

$$
\begin{aligned}
\varphi(2, 2r - 1) &= \sum_{s=1}^{2r-1} \left(e^{2\pi i\left(\frac{2}{2r-1}\right)} \right)^{s^2} \\
&= \sum_{s=1}^{2r-1} \left(e^{i\pi[4/(2r-1)]} \right)^{s^2} \\
&= \sqrt{\frac{i(2r - 1)}{4}} \sum_{s=1}^{4} \left(e^{-i\pi[(2r-1)/4]} \right)^{s^2}.
\end{aligned}
$$

Here we use quadratic reciprocity in the form

$$g(m, n) = \sqrt{\frac{in}{m}}g(-n, m) \text{ if } mn \equiv 0 \pmod{2},$$

where

$$g(m, n) = \sum_{s=1}^{|n|} e^{i\pi \, ms^2/n}.$$

Thus, $\varphi(2, 2r - 1) = \sqrt{\frac{i(2r - 1)}{4}} \sum_{s=0}^{3} e^{i\pi\Gamma s^2}$ where $\Gamma = (1 - 2r)/4$, and

$$
\begin{aligned}
\sum_{s=0}^{3} e^{i\pi\Gamma s^2} &= 1 + e^{i\pi\Gamma} + e^{i\pi 4\Gamma} + e^{i\pi 9\Gamma} \\
&= 1 + e^{i\pi\left(\frac{1-2r}{4}\right)} + e^{i\pi(1 - 2r)} + e^{i\pi\left(\frac{1-2r}{4}\right)} \cdot 9 \\
&= 1 + e^{i\pi\left(\frac{1-2r}{4}\right)} - 1 + e^{2\pi i(1 - 2r)} \cdot e^{i\pi\left(\frac{1-2r}{4}\right)} \\
&= 2e^{i\pi\left(\frac{1-2r}{4}\right)}.
\end{aligned}
$$

Therefore, $\varphi(2, 2r - 1) = \sqrt{i(2r-1)}e^{i\pi\left(\frac{1-2r}{4}\right)}$

$$\varphi(2, 2r - 1) = \sqrt{2r - 1}e^{i\pi((1-r)/2)}.$$

Finally,
$$
\begin{aligned}
\Phi &= \frac{(2(1+i)\sqrt{r(2r+1)})(\sqrt{2r-1}e^{i\pi((1-r)/2)})}{i\sqrt{(2r+1)(2r-1)}} \\
&= 2\sqrt{2}e^{i\pi/4}(-i)\sqrt{r}e^{i\pi((1-r)/2)} \\
&= 2\sqrt{2r}e^{i\pi/4}e^{-i\pi/2}e^{i\pi[(1-r)/2]} \\
&= 2\sqrt{2r}e^{i\pi[(1-2r)/4]} \\
&= (-i)^r 2\sqrt{2r}e^{i\pi/4}.
\end{aligned}
$$

This completes the proof of the claim, and hence the proof of the Lemma as well. ∎

We can now summarize the normalization as follows.

Theorem 8 *Let $A = e^{i\pi/2r}$ and let $|K|$ denote the number of components in a link diagram K. Let $n(K) = b_+(K) - b_-(K)$ be defined as above. Let*

$$Z_K = Z(M^3(K)) = <w*K> \mu^{|K|+1}\alpha^{-n(K)}$$

where

$$\mu = \sqrt{\frac{2}{r}}\,\sin(\pi/r)$$

and

$$\alpha = (-i)^{r-2}e^{i\pi\left[\frac{3(r-2)}{4r}\right]}.$$

Then $Z(M^3(K))$ is an invariant of the 3-manifold $M^3(K)$ and it is normalized so that $Z(S^3) = \sqrt{\frac{2}{r}}\,\sin(\pi/r)$ and $Z(S^2 \times S^1) = 1$.
The proof has been given by the preceding discussion.

12.9   Examples

The Lens Spaces $L_{n,1}$ The next candidate for computation is the lens space $L_{n,1}$. This has surgery presentation an unknotted curve with framing number n.

Let $\Lambda_n = \left\langle \vcenter{\hbox{}} \right\rangle$. Thus, $\Lambda_n = \sum_{a=0}^{r-2} \Delta_a^2 (-1)^{na} A^{na(a+2)}$,

and when $A = e^{i\pi/2r}$ we have $\Delta_a = (-1)^a \sin((a+1)\pi)/\sin(\pi/r)$. From this it is easy to see that

$$\Lambda_n = \frac{1}{\sin^2(\pi/r)} \sum_{a=1}^{r-1} \sin^2(\frac{a\pi}{r})(-1)^{n(a+1)} e^{\frac{i\pi(a^2-1)n}{2r}} .$$

Hence

$$\begin{aligned} Z(L(u,1)) &= \mu^2 \alpha^{-1} \Lambda_n \\ &= (\tfrac{2}{r}) \sin^2(\pi/r) i^{r-2} e^{-i\pi[\frac{3(r-2)}{4r}]} \Lambda_n \end{aligned}$$

Therefore,

$$Z(L(n,1)) = (\frac{2}{r}) i^{r-2} e^{-i\pi\left[\frac{3(r-2)}{4r}\right]} \sum_{a=1}^{r-1} \sin^2(\frac{a\pi}{r})(-1)^{n(a+1)} e^{\frac{i\pi(a^2-1)n}{2r}} .$$

Due to our choice of roots of unity this Lens space evaluation differs from some published versions. The reader should compare it with the formula given by Freed and Gompf [FG91], who computed Lens spaces via Witten's topological quantum field theory. We find that our formula and theirs differ by a power of i. (See also [Nei92] for more sustained bracket calculations of 3-manifold invariants.)

$S^2 \times S^1$. This manifold has surgery datum an unknotted circle with framing number zero. Hence

$$\begin{aligned} Z(S^2 \times S^1) &= \left\langle \vcenter{\hbox{}} \right\rangle \mu^2 \\ &= \mu^{-2}\mu^2 = 1. \end{aligned}$$

This verifies our claim that Z is normalized to 1 on $S^2 \times S^1$.

12.10   Shadow Interpretation

We now apply the work of Section 11 to the 3-manifold invariant $Z(M^3(K))$. Since $Z(M^3(K)) = < w * K > \mu^{|K|+1} \alpha^{-n(K)}$ we can apply Theorem 5 to the evaluation of $< w * K >$ as a shadow-world partition function. Since

$< w * K >$ assigns the sum $\sum_i \Delta_i \, \underline{\underline{\quad}}^{\,i} = w$ to each component of K,

we see at once that the shadow-world partition function for $< w * K >$ can regarded as the partition function associated with a two-cell complex obtained from the two-cell complex for the shadow of K (on S^2 or $\mathbb{R}^2$) by attaching one 2-cell along each component of K. The Δ_i-summation corresponding to w for each component is then catalogued by the weights Δ_i assigned to the 2-cells for each component. This gives a quite natural expression for $Z(M^3(K))$ as the partition function for this two-cell complex. This approach has been used to compute all the numerical examples of Chapter 14.

Formally, we can state the

Theorem 9 *Let K be a link diagram, regarded as surgery data for the 3-manifold $M^3(K)$. Let $Z(M^3(K))$ be the Witten-Reshetikhin-Turaev invariant as described above. Let $Sh(K)$ denote the corresponding shadow diagram for K with the caveat that a 2-cell is attached to each link component. Assign weights to each coloring of the faces of $Sh(K)$ from the set $\{0, 1, 2, \ldots, r-2\}$ by associating Δ_a to a face colored a, $\theta(a, b, c)$ to an edge touching regions labelled a, b, c and modified tetrahedral symbols*

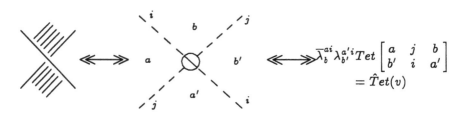

(invert $\lambda's$ for reverse crossing) to the shadow crossings. Let $Z_0(Sh(K))$ denote the partition function obtained from these weights by the formula

$$Z_0(Sh(K)) = \mu^{|K|+1} \alpha^{-n(K)} \sum_c \prod_{v,e,f} \frac{\hat{T}et(v; c) \Delta_{f,c}^{\chi(f)}}{\theta(e; c)^{\chi(e)}}$$

where $\hat{T}et(v; c)$ denotes the modified tetrahedral symbol for the vertex v and coloring c of $Sh(K)$, $\Delta_{f,c}^{\chi(f)} = \Delta_i^{\chi(f)}$ if c assigns i of and $\theta(e; c)$ denotes the θ-evaluation assigned to the edge e. ($\chi(e) = 0$ if e has no nodes and 1 if *e does have nodes). The sum is over all q-admissible ($q = e^{i\pi/r}$) colorings with the unbounded region of the diagram receiving color 0 (hence Z_0). The product is over all vertices, edges and faces of $Sh(K)$. The factors μ and α are as defined for the invariant $Z(M^3(K))$. Then*

$$Z(M^3(K)) = Z_0(Sh(K)).$$

Proof: This follows at once from the constructions of the previous chapter, plus the fact that

$$Z(M^3(K)) = \mu^{|K|+1}\alpha^{-n(K)} \sum_{\vec{a}} \Delta_{a_1}\Delta_{a_2}, \cdots \Delta_{a_n} < \vec{a} * K >$$

where the $a_i \in \{0, \ldots, r-2\}$ and label the components of K. Once these labels are transfered to the cells bounding each component of $Sh(K)$, rest of the identity follows from our previous shadow translation of $< \vec{a} * K >$. ∎

Concluding Remarks. The applicability of the formula $Z(M^3(K)) = Z_0(Sh(K))$ is an interesting open question. By using the shadow world translation, it does give a direct formula for the WRT invariant in terms of recoupling coefficients. This seems to be the best approach for general numerical computations. On the other hand, the remarkable similarity in form between the partition function $Z_0(Sh(K))$ and the partition function for the Turaev-Viro invariant $TV(M^3)$ is part of a larger picture. The shadow $Sh(K)$ we have used above with one extra cell for each link component can be regarded as a "shadow" of $W^4(K) = \{$the handle-body obtained by attaching a 2-handle to each link component in $S^3 = \partial D^4\}$. In [Tur92] this technique is used in conjunction with the fact that $\partial(W^4(K) \times I) = M^3(K) \times I \cup W^4(K) \times 0 \cup -W^4(K) \times I$ to show that $I(M^4(K)) = Z(M^3(K)) \cdot \overline{Z(M^3(K))}$. The idea of the proof is that an appropriate decomposition of $W^4 \times I$ induces the "internal shadow" of the special spine for $M^3(K)$ and also the "external shadows" $Sh(K), Sh(-K)$ for $W^4(K)$ and $-W^4(K)$. Cobordism properties then yield the product identity $I(M^3) = Z(M^3) \cdot \overline{Z(M^3)}$. Here we have developed a background for these results through the organization of a combinatorial theory of recoupling. Obviously more work remains to be done in this domain.

12.11   Appendix: Invariants of 4-Manifolds

The purpose of this appendix is to give a sketch of Justin Robert's beautiful method [Rob93] of understanding the Turaev-Viro Invariant, and also to discuss the relationship of this approach with the structure of the 4-manifold invariant of Crane and Yetter [CY93]. This combination of methods produces a state summation formula for the signature of a 4-manifold.

The *encirclement lemma* (Chapter 12, Lemma 22 of these notes) in combination with recoupling yields formulas for several colored strands with an omega (ω) belt about them:

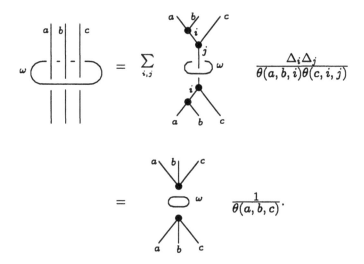

Note that, by Lemma 22, a summand is non-null only when $j = 0$ and thus, $i = c$. In [Rob93], this trick is used to analyse the structure of some invariants of 3-manifolds and 4-manifolds. In the case of 3-manifolds, one can present a 3-manifold by pasting together two handlebodies. (A handlebody is a 3-ball with 1-handles ($D^2 \times [0, 1]$) attached along embeddings of $D^2 \times 0$ and $D^2 \times 1$ in the boundary of the 3-ball.) It is sufficient, to specify such a pasting, to give two sets of curves on a surface S that represents the boundary of either handlebody. The first set of curves is a collection of the basic curves that bound disks in the first handlebody (so that this handlebody is returned to a ball upon being cut along these disks). The second set of curves is in correspondence with the bounding curves for the second handlebody. It is customary to draw the first set of curves as the meridian curves on a standard representation of the surface.

The second set of curves will then have the complexity that corresponds to the structure of the pasting map between the two handlebodies. The resulting picture of two sets of curves on a given surface is called a *Heegaard Diagram* for the 3-manifold.

Now take such a diagram, embed it in 3-space (any way you please). Push each of the meridians in the first set of curves along the outward normal to the surface so that they do not intersect the curves in the second set. Call this link $L(H)$ where H is the original Heegard decomposition. Let $M(H)$ denote the 3-manifold that corresponds to H. The usual normalization of $< \omega^* L(H) >$ is an invariant of the 3-manifold $M(H)$. In fact, Roberts shows that it is the Kauffman-Lins (Chapter 10 of this monograph) version of the Turaev-Viro invariant.

The proof is based on the encirclement recoupling formulas and the geometry of the handlebody decomposition that is obtained from the dual 1-skeleton of a triangulation of $M = M(H)$. In the dual 1-skeleton we have four edges incident to each 0-simplex. The handlebodies consist in a tubular neighborhood of this structure, and its complement.

Thus $L(H)$ looks locally as shown below.

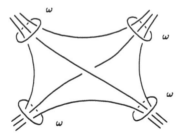

Applying the encirclement lemma yields isolated tetrahedral nets at each vertex.

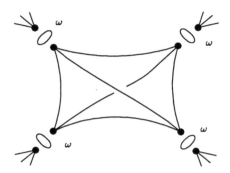

The invariant $< \omega^* L(H) >$ is therefore seen as a sum of products of evaluations of tetrahedral nets and other terms corresponding to the details of the recoupling formulas. It is easy to see that this is exactly the Kauffman-Lins version of the Turaev-Viro invariant. Roberts uses this method to give a simple proof that the Turaev-Viro invariant is the product of the WRT invariant and its complex conjugate. This theorem was first proved independently by Turaev and by Walker.

Roberts then goes on to discuss a four dimensional version of the construction $M \longrightarrow M(H) \longrightarrow L(H) \longrightarrow < \omega^* L(H) >$. The upshot of this description is that one can again compute $< \omega^* L >$ for the link obtained from curves for the 2-handles together with pushoffs for the meridians of the thickened and dimensionally reduced dual 1-skeleton. The story about computing $< \omega^* L >$ is then the same as before except that there are five tubes meeting at a node in this dual thickened skeleton (since we are working with the 4-simplices of triangulation of M). Therefore the encirclement and recoupling works as shown below.

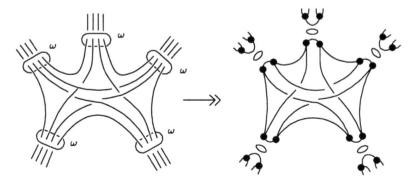

This means that the resulting invariant of the 4-manifold is a sum of products of "15-j" symbols and some other terms.

An analysis of Robert's construction by Crane, Kauffman and Yetter [CKY93] leads to the following theorem:

Theorem 10 *Let W be a closed 4-manifold. Let $\sigma(W)$ denote the signature of W, $\chi(W)$ denote the Euler characteristic of W and $CY(W)$ denote the Crane-Yetter invariant of W for the choice of $A = exp(2\pi i/4r)$ in the bracket evaluations. Let values N and k be defined as shown below, for r a positive integer greater than 2:*

$$N = 2r/(q - q^{-1})^2$$

$$\kappa = exp(i\pi(-3 - r^2)/2r)exp(-i\pi/4).$$

Then $CY(W) = \kappa^{\sigma(W)} N^{\chi(W)/2}$.

One consequence of this Theorem is that for sufficiently large r ($r >$ the number of 2-simplices in a triangulation of W), the signature of W is computed through purely local combinatorial data about W. This unexpectedly answers an old question about the existence of a local combinatorial formula for the signature of a 4-manifold.

Chapter 13

Blinks ↦ 3-Gems: Recognizing 3-Manifolds

13.1 Motivating 3-Gems

The possibility of obtaining the Witten-Reshetikhin-Turaev invariants from blackboard framed links studied in the last Chapter makes possible effective calculations of this strong family of invariants for 3-manifolds. It is known that these invariants are not complete invariants. However, beyond opening the ways for theoretical explorations, they also help in a project of explicit listing of 'small' 3-manifolds without duplications nor ommissions: they suggest possible duplicates, which are to be proven so by other methods. One difficulty with the presentation by framed links is that the recognition of 3-manifolds from them is very difficult, due to the lack of useful simplification procedures.

The purpose of this Chapter is to provide connections between blackboard framed links and 3-*gems*, objects which permit the precise topological recognition of a great part of the small 3-manifolds obtained from these links, as can be noted from the tables in Chapter 14. We also review basic facts on 3-gems to make the exposition self-contained.

A drawing of a projection of a link can be checkerboard colored: the outer region is white, their neigboors by an edge are colored black and so on in a way that every edge separates a black from a white region. *The graph of a projection of a link* is the plane graph which has one vertex inside

each black region and which has an edge for each crossing of the projection
so that the edge bissects the corresponding crossing. A crossing of a link
projection is a *slash crossing* if its overpass crosses the corresponding edge
of its graph from northeast to southwest. Otherwise, it goes from northwest
to southeast and is a *backslash crossing* as below:

A *blink* is a finite plane graph with a bipartition in the edges. In the
diagrams this bipartition is given by plain and dotted edges. There is a
1-1 correspondence between blinks and blackboard framed links: from a
blink draw its medial graph and adjust crossings so that slash crossings
correspond to plain edges and backslash crossings correspond to dotted
edges. Clearly, the construction is reversible. Here is an example of this
1-1 correspondence, involving Walle's link.[1]

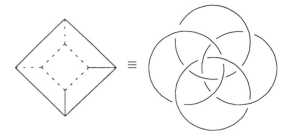

Blinks, via their blackboarded framed links, are objects which encode
orientable connected 3-manifolds in a very concise way. Moreover, they are
the minimal data structure required to get the Witten-Reshetikhin-Turaev
invariants: indeed, the variables associated with the partition function to
compute these invariant are in $1 - 1$ correpondence with the vertices, faces
and *zigzags*[2] of the blink.

---

[1]This link is credited by Stewart [Ste89] to C. van de Walle with the claim that
removing any component frees the other three. But this is true only for two of the
components: dropping one component of the other two the result is the borromean link.

[2]A zigzag is a closed path that alternates turning to the rightmost and to the left-
most edge at each vertex. There are no abstract differences, between stars of vertices,
boundaries of faces or zigzag cycles of a graph embedded in a closed surface [Lin82].

As we show in Section 5, from a connected blink we can get a special 4-regular 4-edge colored graph, named a 3-gem, which encodes the same orientable connected 3-manifold. There is a set of two simple moves *(dipole moves)* on 3-gems which were proven by Ferri and Gagliardy (in the context of crystallizations) to suffice for generating homeomorphisms. As a consequence of their Theorem, we provide, in Section 4, an alternative simple proof of the sufficiency of the Matveev-Piergallini moves for the invariant of Chapter 10 (Turaev-Viro invariant).

The orientable 3-manifolds representable by 3-gems with less than 30 vertices have been completely classified [LD89], [LD91]. From the algorithms developed for this classification we recognize, from blinks, various well known 3-manifolds for which we display tables of Witten-Reshetikhin-Turaev invariants in the next chapter.

The basis for the 3-gem classification is a function named $TS\rho$-*algorithm* which takes a 3-gem and maps it into a class of reduced 3-gems. The 3-gem model and the basic facts on the classification are discussed starting from the next section. With $TS\rho$ we identify various blinks inducing the same 3-manifold. For instance, here are the first Witten-Reshetikhin-Turaev invariants computed at $A = e^{i\pi/2r}$ (with the two normalizations discussed in Chapter 12) for the 3-manifold induced by Walle's blink:

r	$S^1 \times S^2 - normalization$	$S^3 - normalization$	$C^{++}, 486$ time	$r-$ states
3	+2.82842712	+4.00000000	0.093sec	16
4	+6.00000000	+12.00000000	1.885sec	192
5	+10.75997619	+28.94427191	21.366sec	1856
6	+17.32050808	+60.00000000	152.144sec	13955
7	+25.87092151	+111.55076332	789.667sec	83664

These numbers agree with the ones for $S^1 \times S^1 \times S^1 \# S^2 \times S^1$. After noting this fact we used $TS\rho$ on the 3-gems induced by the two blackboard framed links below to arrive to the same 3-manifold: $R_{24}(1)$ with handle number 1.[3]

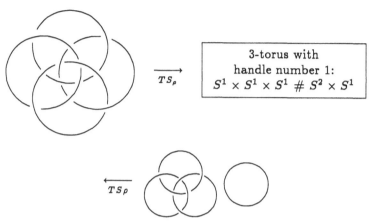

Note that the second blackboard framed link above is the smallest framed link inducing $S^1 \times S^1 \times S^1 \,\#\, S^2 \times S^1$. We note also that $R_{24}(1)$ is a notation to mean the first rigid 3-gem with 24 vertices [LD89]. This is the single 3-gem with 24 vertices which induces the 3-torus.

We note that an important characteristic of the 3-gem model is the possibility of quick solution for the isomorphism problem, via the *code*, based on the *rooted numbering algorithm*, which is treated in Section 7.

The last Section, Section 8, provide the TS-moves, which are local moves on 3-gems basic for their useful simplification theory.

Sections 13.2 — 13.3 and 13.6 — 13.8 appear originally in [LS92]. They are included here with a few changes to settle the terminology on 3-gems and make our presentation self-contained.

13.2   Graph-Encoded 3-Manifolds

An $(n + 1)$-*graph* is a finite graph G where at each vertex meet $n + 1$ differently colored edges; each edge has a color of a total of $n+1$ colors. An m-*residue* of G is a connected component of a subgraph generated by any m specified colors. Note that the 2-residues are even sided bicolored polygons in G, also named *bigons*. Let G^2 be the 2-complex obtained from G by attaching a 2-cell to each bigon. A *3-dimensional graph encoded manifold*

---

[3]It seems difficult to show that these two blackboard framed links induce the same 3-manifold by the moves of Chapter 12.

or simply a 3-*gem* is a $(3+1)$-graph G satisfying the following arithmetical condition:

$$v + t = b,$$

where v, t, b stand for the number of $0-, 3-, 2-$residues of G, respectively. This arithmetical condition holds if and only if each 3-residue with its attached 2-cells forms a topological 2-sphere [Lin88]. If G is a 3-gem, let G^3 be the 3-complex obtained from G^2 by attaching a distinct 3-cell to the 2-sphere associated to each 3-residue.

It is easy to show that if G is a 3-gem, then the associated topological space $|G| = |G^3|$ is a closed 3-manifold. We leave this verification to the reader. Reciprocally, we have

Proposition 13 *Let M^3 be any closed 3-manifold. Then there is a 3-gem G so that $|G| \cong M^3$.*

Proof: This is an easy consequence of Moise's triangulation theorem for 3-manifolds [Moi77]. Let T be a triangulation for M^3 and T' its barycentric subdivision. The vertices (0-cells) of T' are naturally partitioned into four classes V_0, V_1, V_2, V_3. The class V_0 is formed by the original vertices of T. For $i = 1, 2, 3$, the class V_i is formed by those vertices which are in 1–1 correspondence with the i-cells of T, situated in the barycentre of an i-cell. Note that each tetrahedron of T' has the four vertices in distinct classes. This 4-partition of the vertices of T' induces a 4-coloration of its triangular 2-cells: color with i the face of a tetrahedron in T' whose opposite vertex is in V_i.

Let G be the 1-skeleton of the cell decomposition dual to T'. The 4-coloration of the 2-cells of T' is inherited, under geometric duality, by the edges (1-cells) of G. Graph G with the edges so colored is a 3-gem and satisfies $|G| \cong M^3$.   ∎

If a closed 3-manifold M^3 and a 3-gem G satisfy $|G| \cong M^3$, then we say that G *represents* M^3. The 3-gem obtained in the above Proposition is much bigger than it needs to be in the sense that there are *smaller* (with fewer vertices) 3-gems representing the same M^3. One point in favor of the representation by 3-gems is that they have a rich simplification theory. At the bottom there appear highly structured graphs, where uniqueness is not unfrequent.

We are concerned only with **orientable** manifolds. The orientability manifests itself as the bipartiteness of the representing gems [LM85]. Therefore an orientation is specified by a bipartition of the vertices into two ordered classes. In the pictures, the bipartition is sometimes depicted by drawing small full and hollow circles on the vertices. We also use odd-even numbers and lower-upper case letters to present this bipartition.

In order to achieve compacteness, our diagrams for 3-gems obey various conventions. We use labels for the vertices. The edges of color 0 are represented implicitly by a pair of labels of type $(2n-1, 2n)$, or by a pair of lower, upper case letters, (x, X). The color of an edge is given by number of marks on it. For instance, here is a (canonical) presentation of the unique minimum 3-gem representing the 3-manifold S^3/Q_8.

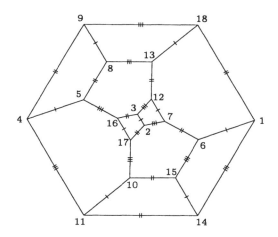

The vertex labelling of a canonical presentation is very special in the sense that it can be recovered from the 3-gem, by the algorithm in Section 7.

In small 3-gems, it is sometimes convenient to replace the numerical vertex-labellings 1, 2, 3, 4,...by literal ones a, A, b, B,...The four 3-residues of the above 3-gem with literal labels are

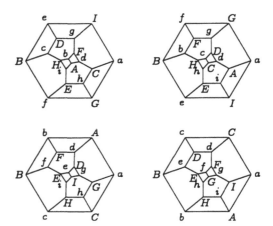

The first one coincides with the 3-residue not involving color 0. The other three are obtained by leaving out, respectively, colors 1, 2 and 3. The fact that these residues are isomorphic is not always true, but is frequent in minimum 3-gems for common 3-manifolds. The manifold S^3/Q_8 is obtained from the four balls above by pasting together pairs of faces in such a way as to identify equally labelled vertices. From the vertex labelled 0-missing 3-residue we can recover the 3-gem: note that color 0 is given by the pairing and the boundaries of the faces are bicolored; this fact yields the other three colors up to permutations.

We define the *gem-complexity* of a 3-manifold M^3, $\xi(M^3)$, as one less than half of the number of vertices in a 3-gem representing M^3 and having the minimum number of vertices among all 3-gems representing it. For example, the complexity of S^3/Q_8 is 8 which is attained by the above 3-gem with 18 vertices.

All the 3-manifolds up to gem-complexity 13 have been topologically classified, [LD89], [LD91]. The rigid 3-gems of with 30 vertices have been generated and $TS\rho$-classified, [Dur92]: the 56912 rigid 3-gems with 30 vertices produce only 56 essential TS-classes (not counting gems which induce non-trivial connected sums). The topological classification is under way.

There are finitely many 3-manifolds of a given gem-complexity. Moreover we conjecture that this function is additive on connected sums.[4] A similar complexity for 3-manifolds was introduced by Matveev and has the

---

[4] This conjecture is the reason to subtract one in the definition of complexity.

finitiness and the additive property [Mat91]. If one proves that the complexity ξ is additive on connected sums then we could avoid appealing to the truth of the Poincaré Conjecture to guarantee that the *prime gems* (the ones not displaying a non-separating 2-sphere) in the catalogue [LD89] represent indeed prime 3-manifolds.

13.3   Dipole Moves: Ferri-Gagliardi Theorem

The generic 3-gem obtained in Proposition 1 is much bigger than the minimal 3-gem representing the same M^3. The class of minimum[5] 3-gems representing a specific 3-manifold are formed by highly structured graphs and often they have only one member [LD89], [LD91], [Dur92].

The primitive simplifying sub-configuration in a 3-gem is the "*k*-dipole" which we now explain. Let the colors attached to the edges of a 3-gem G be labelled $0, 1, 2, 3$. Two vertices of G linked by k-edges ($k = 0, 1, 2, 3$) whose color set is K constitute a *k-dipole* if they are in distinct components of the subgraph generated by all edges with colors in $\{0, 1, 2, 3\} \backslash K$.

The *cancellation of a k-dipole* is the following operation: remove the two vertices and all the edges linking them; this gives $2 \times (4 - k)$ pendant edges; identify pairwise the ends incident to edges of the same color. The inverse operation is called a *k-dipole creation*. A *k-dipole move* is either its cancellation or creation. The diagrams below display dipole moves for $k = 0, 1, 2, 3$. The color of the edges are indicated by the number of marks on them. Clearly, the moves are indicated up to color permutation and the diagrams for them are not entirely local: we are assuming in the complement of the diagrams connectivity restrictions which characterize dipoles.

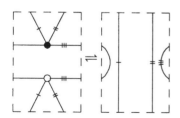   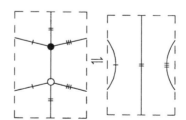

---

[5]Having smallest possible number of vertices

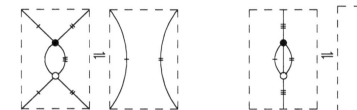

It is not difficult to observe that the k-dipole moves for $k = 1, 2, 3$ do not change the represented manifolds. As for the cancellation of a 0-dipole, it corresponds to the attachement of a handle [Lin85].

A basic result of the theory is the following Theorem:

Theorem 11 (Ferri and Gagliardi [FG82]) *If M^3 and N^3 are homeomorphic 3-manifolds, then any 3-gem G_M representing M^3 can be transformed into any 3-gem G_N representing N^3 by means of 2- and 1-dipole moves.*

This Theorem is proved for general dimensions and is stated in terms of *crystallizations*, which can be defined as a gem without 1-dipoles.

13.4   The   Sufficiency   of   the   Matveev-Piergallini Moves

To exemplify an application for the Ferri-Gagliardi Theorem we use it to give a simple proof that the invariance of I_q (see Chapter 10) depends only on the invariance under the Matveev-Piergallini moves II and III, and that it behaves well under edge dilation.

Note first that the 2-skeleton of a 3-gem is a special spine as defined in Chapter 10. The presence of 3-dipole implies the presence of two 1-dipoles each having one vertex in common with the 3-dipole. Cancelling the 3-dipole is accomplished by cancelling any one of the companion 1-dipoles. Note also that in the context of 3-gems, 2-dipole creation is a particular case of move II (lune move) of Chapter 10. Therefore, it is enough to show the invariance of I_q under the cancellation of a 1-dipole. In doing so we leave momemtarily the class of 3-gems but remain all the time in the class of special spines (see Chapter 10). The proof of the main Lemma below is surprisingly easy.

Lemma 24 (Factorization of 1-dipole cancellation) *The cancellation of a 1 dipole in a 3-gem is factorable as a finite sequence of lune moves, Y-moves, their inverses and one 3-dipole cancellation.*

Proof: Let α be an 1-dipole in a 3-gem. Instead of eliminating α we perform an Y-move on it. This requires an explanation. The picture that we have used to define Matveev-Piergallini's Y-move is adequate to derive the invariance of TV_q under it. However, the picture is not the most symmetric one. The move is in fact dependent only on the edge and independs on which one of six possible 2-cells is pushed through. Here is an isotopic deformation of the defining picture of the Y-move. It proves automatically this independence, via the symmetry:

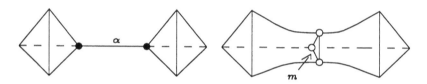

There are nine 2-cells present in the left picture above: three incident to an end of α and not to the other, three more with this property with the roles of the ends exchanged, and three incident to α. Thus, there are ten 2-cells present on the right picture. Therefore, this move is defined by the "edge α of a special 2-complex which becomes a triangular region m". In the context of 3-gems, if the edge α is a 1-dipole, then the region m separates the (distinct) 3-cells which correspond to the (distinct) 3-residues of the ends of α.

What we have to do in order to accomplish the 1-dipole elimination is to free ourselves of the region m. Think of it as a membrane, which we start pushing so as to contract its boundary by applying Y-moves, lune moves and its inverses. It is easy to arrive at a situation where the boundary of m crosses twice a unique edge of the 3-gem. At this moment we have at hand a dilated edge. For instance, here is this process in a particular case:

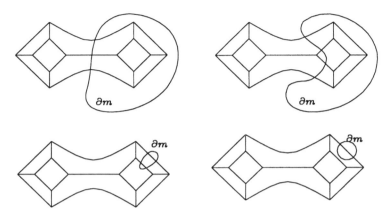

The graph is the resulting 3-residue and is supposed embedded into a 2-sphere (which is the boundary of the 3-cell U corresponding to the 3-residue) and the curve corresponds to the boundary of the membrane m which is (except for the boundary ∂m) embedded inside U and separates it into two 3-cells. We present the first step (an Y-move) and the last step (a lune move) of a possible procedure of this kind. Observe that after obtaining the dilated edge we have a 3-gem again. The dilated edge is a 3-dipole whose cancellation completes the proof.   ∎

Theorem 12 *Let M^3 be a closed 3-manifold and G be a 3-gem inducing M^3. Let t be the number of 3-residues of G. Then,*

$$I_q(|G|) = \tau^{-(t-1)}TV_{|G|,q}$$

is a topological invariant.

Proof: The 2-dipole creation is the lune move; t and $TV_{|G|,q}$ are invariant and so is I_q. A 1-dipole cancellation divides $TV_{|G|,q}$ by τ; but t decreases by one. Whence, I_q is invariant under 1-dipole cancellation (or creation). Therefore, the topological invariance follows from the Ferri-Gagliardi Theorem.   ∎

13.5   From Blinks to 3-Gems

Suppose we are given a blink. Draw the associated blackboard framed link. Then it is rather easy to construct a Heegaard diagram yielding the same manifold as the one given by the surgery presentation. Consider each component in the surface of a thin solid torus. At each crossing make a bridge linking upper and lower tori with a solid cylinder, as shown in the picture below.

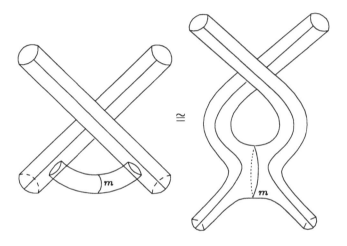

After the addition of one cylinder for each crossing we get a handlebody.

Proposition 14 *The handlebody above constructed and the original link components embedded on it together with the meridian curves of the added cylinders form a Heegaard diagram for the same manifold given by the link with blackboard framing.*

Proof: The proof is straightforward from the definitions of surgery on framed links and of Heegaard diagrams.   ■

We can indicate the passage from the projection of a blackboard framed link to a Heegaard diagram by repeating at each crossing the local substitution depicted below:

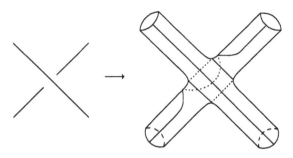

Here is an example for S^3 given by surgery on  · The associated handlebody has genus two and the Heegaard diagram in this simple case is:

To obtain a 3-gem representing the same 3-manifold as the one represented by a Heegaard diagram is simple: start by drawing a parallel curve to each curve in the Heegaard diagram. Each vertex of the 3-gem is defined from a transition of a curve from an arc behind to one in front. Let the behind arcs be edges of color 3 and frontal arcs be edges of color 0. There are now an even number of vertices around each hole. Color the boundary of each such hole alternatively with colors 1 and 2, in such a way that the 3-residue not involving color 3 has bicolored boundaries of faces.

It is straightforward to prove that the resulting 4-regular graph is a 3-gem inducing the same 3-manifold as the one given by the Heegaard diagram. We leave the details for the reader. In the above example, after

the doubling we get

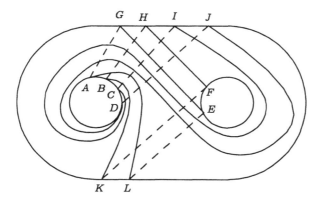

This produces the 3-gem

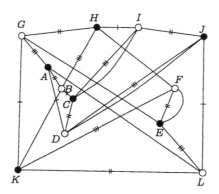

A sequence of dipole cancellations get us to the canonical 3-gem for S^3: the one with two vertices.

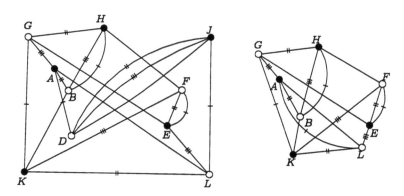

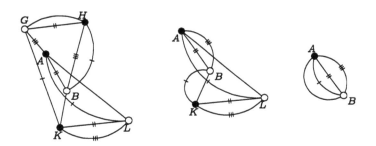

From the above discussion it is straightforward, in general, to go from a blackboard framed link with n crossings to a 3-gem with $12n$ vertices inducing the same 3-manifold. Each crossing is replaced by a fixed partial

3-gem, as shown below:

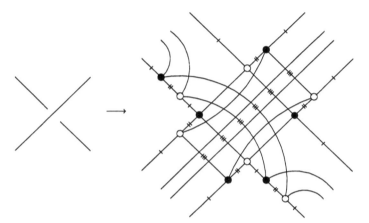

The algorithm realizing the transformation *blink* ↦ 3-*gem* has been implemented and is used to recognize the 3-manifolds induced by simple blinks. The 3-gems have a richer simplification theory implying that it is easier to recognize 3-manifolds given by small 3-gems than by small blinks. Therefore, the above *"multiplication by 12"* pays off.

There are two observations about the above transformation: (i) blinks can only induce connected 3-manifolds. A disconnected blink induces the 3-manifold given by the connected sum of the disjoint 3-manifolds given by each connected component of the blink. Thus, we need to ambient isotope the link inducing the blink so it becames connected. (ii) if the blink is a single vertex, there is no way to perform the above transformation. We need to deform the inducing circle to get crossings in the projection. Indeed, from the second blackboard framed link below we get a 3-gem for $S^1 \times S^2$.

13.6   Rigid 3-Gems

Once a k-dipole is discovered in a 3-gem, a smaller 3-gem representing the same manifold (up to handles in the case of 0-dipole) is obtainable by dipole cancellation. Thus, the presence of dipoles permits a straightforward reduction. Another reducing configuration is the ρ-pair. A ρ-*pair* corresponds to two edges having the same color and which are shared by two or three bigons.

The *switching of a ρ-pair* is the deletion of its constituent edges followed by the introduction of another pair of the same color, with the same ends, so as to preserve the bipartition.

Proposition 15 (Ferri and Lins [FL91]) *Let the bipartite 3-gem G have a ρ-pair and let G' be the resulting bipartite $(3+1)$-graph obtained by switching the pair. Then*

1. *G' is a 3-gem;*

2. *G' has a 1-dipole or a 0-dipole;*

3. *$|G'| \cong |G|$ or $|G'| \cong |G| \# S^1 \times S^2$ or $|G'| \cong |G_1| \# |G_2|$, where the disjoint union of G_1 and G_2 is G'.*

From the second part of the above Theorem, it follows that a ρ-pair induces a smaller 3-gem representing, up to handles, the same manifold. A ρ-*move* on a ρ-pair is its switching followed by the cancellation of a 0-dipoles or of 1-dipole(s). In the catalogue of [LD89] we do not keep 3-gems with ρ-pairs, since they are irrelevant and unecessary for extending the catalogue.

A 3-gem is *rigid* if it is bipartite and free of both dipoles and ρ-pairs. It is a fact that 3-residues of such gems are 3-connected. By a Theorem of Whitney, 3-connected graphs have a unique embedding in the 2-sphere, up to orientation. From the existence of rigid 3-gems representing S^3 and $S^2 \times S^1$ it is easy to establish that

Proposition 16 (Lins and Durand [LD89]) *Any closed orientable 3-manifold is representable by a rigid 3-gem.*

A 3-residue of a rigid 3-gem is a bipartite cubic 3-connected graph embeddable in a 2-sphere. The embedding so that the faces are the bigons is unique up to orientation. We call such embedded graphs as 3*-*graphs*.

Fusion of an edge of a 3*-graph with at least four vertices is the process of deletion of the edge together with its ends followed by the pairwise identification of the four free ends so that a bipartite cubic graph embedded into the 2-sphere results.

If the resulting graph is 3-connected, i.e., the fusion is internal to the class of 3*-graphs, then the edge is called a *fusible* edge.

The 1-skeleton of a prism whose base polygons have an even number, $2n, n \geq 2$, of sides, form a simple family of 3*-graphs, denoted by P_{2n}.

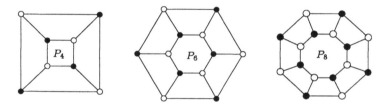

Note that no edge of P_{2n} is fusible. However, the following Theorem holds and it provides a recursive generation of all 3*-graphs with $2n + 2$ vertices from the ones with $2n$ vertices.

Proposition 17 (Lins [Lin84]) *A 3*-graph G has a fusible edge or else it is isomorphic to P_{2n}, for some $n \geq 2$.*

Each 3*-graph H is (up to color permutation) uniquely 3-edge colorable. To all 3-gems having H as one of its 3-residues, we have to extend it by specifying new edges of a fourth color forming a perfect matching. By trying all possible extensions of all 3*-graphs with $2n$ vertices and keeping the ones which provide rigid 3-gems, all rigid 3-gems with $2n$ vertices will be eventually generated. This strategy has been implemented in [LD89]. An important tool which makes the strategy possible is that the isomorphism problem for $(3 + 1)$-graphs is computationally tractable, as we show in the next Section.

13.7   The Code of a Bipartite $(n+1)$-Graph

Let $\Delta_n = \{0, 1, 2, \ldots, n\}$ be the set of colors attached to the edges of an $(n+1)$-graph G. Given a permutation π of Δ_n and a vertex r of G we define a bijection:

$$N_\pi^r : V(G) \longrightarrow \{1, 2, \ldots, |V(G)|\}$$

by a standard algorithm. This algorithm has complexity proportional to the number of edges of G and is named the *rooted numbering algorithm*. It assumes that G is connected and that it is described in terms of an $(n+1) \times |V(G)|$ matrix also denoted by G. The algorithm also uses a *stack* S of even numbered vertices and a set U of used odd numbered vertices. The variable v indicates the current vertex and the variable i is used to assign a unique number to the vertices. Each even numbered vertex is pushed into S and popped out from S exactly once. The boolean variable found is used to mark the discovery of the next odd numbered vertex. Let $G(c, v)$ be the neighbor of vertex v by the c-colored edge.

The Rooted Numbering Algorithm

```
(Initialization)
S ⟵ ∅; U ⟵ ∅; i ⟵ 1; v ⟵ r; N_π^r(r) ⟵ 1;
(Main loop)
repeat
    U ⟵ U ∪ {v};
    v ⟵ G(π(0), v);
    N_π^r(v) ⟵ 2i; push(S, v);
    found ⟵ false;
    repeat
        if {G(π(c), v)|1 ≤ c ≤ n}\U ≠ ∅ then begin
                c ⟵ min{π(c)|G(π(c), v) ∉ U};
                v ⟵ G(c, v); N_π^r(v) ⟵ 2i + 1;
                found ⟵ true;
        end
        else if not(S = ∅) then v ⟵ pop(S);
    until flag or (S = ∅);
    if flag then i ⟵ i + 1;
until (S = ∅).
```

Note that we are assuming G to be connected, each vertex gets a distinct number of $\{1, 2, \ldots, |V(G)|\}$. The N_π^r-code for G is the $(n \times \frac{1}{2}|V(G)|)$ matrix whose (i, j)-entry is

$$\frac{1}{2} N_\pi^r(G(\pi(i), (N_\pi^r)^{-1}(2j - 1))).$$

Here, $1 \leq i \leq n$ and $1 \leq j \leq \frac{1}{2}|V(G)|$. Thus, the neighbors by color $\pi(0)$ are not explicitly given by the matrix. However, we know that the number of the $\pi(0)$-neighbor of vertex number i is $i + 1$ if i is odd and $i - 1$ if i is even. Also we take advantage of the bipartiteness and present explicitly only the neighbors of the odd numbered vertices. These are even numbers and we use such numbers divided by two in the codes.

Observe that from any of its N_π^r-codes, we can recover G up to color permutation. In fact, the permutation π is lost and we *canonically* color the edges by assuming the permutation to be the identity.

The *code for a bipartite $(n + 1)$-graph G* is the lexicographic maximum among the N_π^r-codes for G, where we write the rows in sequence and consider each N_π^r-code as a single vector with $\frac{1}{2}n|V(G)|$ coordinates.

The vertex labelling of the canonical presentations which we use all attain the code of the corresponding 3-gem. We also replace the numerical labellings given by the rooted numbering algorithm by the literal ones, as explained before. Instead of dividing by two we replace the upper letters by lower case ones. Thus the code of the 3-gem representing S^3/Q_8 is read directly from its canonical presentation:

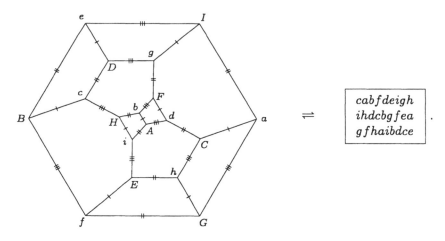

There is a $1 - 1$ correpondence between codes and canonical presentations of 3-gems.

Theorem 13 *Two connected bipartite $(n+1)$-graphs are isomorphic up to color permutation if and only if their codes coincide.*

Proof: This is a straighforward consequence of the rooted numbering algorithm which permits defining the N_π^r-codes and from the definition of the code as the lexicographic maximum.   ∎

13.8   TS-Classes: a Basis for 3-Manifold Classification

Let G be a 3-gem and $B(G)$ the set of its bigons. For $a \in B(G)$ let s_a denote the the numer of edges in the bigon a. Let also t_G be the number of 3-residues of G.

Proposition 18 (Lemma 2 of [Lin86]) *For any 3-gem G the following equality holds:*

$$\sum_{a \in B(G)} (3 - \frac{1}{2}s_a) = 3 \times t_G.$$

The above Proposition implies that a 3-gem without multiple edges with the same ends, (like the rigid ones) must contain at least 12 square bigons. If there are bigons a with $s_a \geq 6$, the number of square bigons increases to compensate for the negative summands that arise. Thus, a rigid 3-gem must have many square bigons. This explain the frequent appearance of the configurations below formed by three square bigons, and are called, for this reason TS-configurations:

 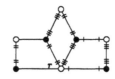

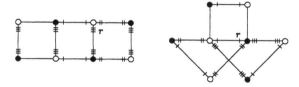

These configurations, their *root* vertex r and their fixed colorings are used as referencials for keeping track of moves in the classification [LD91]. A TS-configuration is specified in a canonically labelled 3-gem by giving its type, the number of the root vertex r and the a permutation. This permutation is read by comparing the canonical colors of the edges induced by the code against one of the above colored configurations. Indeed, the specific colors shown in the above configurations are, by an arbitrary but fixed convention, used as reference.

The TS-configurations imply six involutions named TS-moves. These moves are factorable into dipole moves [LD91] and so maintain the represented space. However they may cause the appearance of dipoles or ρ-pairs which yields smaller 3-gems. This is a useful technique for simplifying these objects.

In order to describe these moves geometrically we must draw properly the outgoing edges of a TS-configuration.

The first configuration provides three TS-moves, which we define below, up to color permutation. The first TS-move consists of a local π-rotation and a black/white exchange of the vertices:

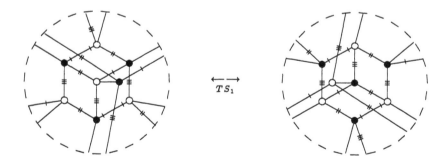

The second TS-move consists of a local reflection in the horizontal diameter:

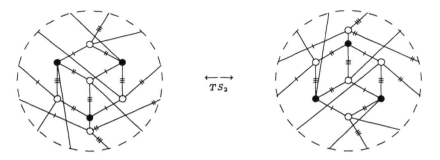

The third TS-move consists of a local reflection in the horizontal diameter and a local black/white vertex reversal:

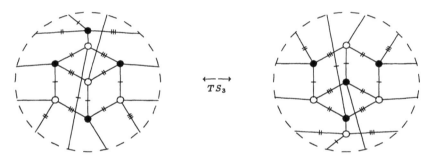

The fourth TS-move consists of a local reflection in the horizontal diameter, a local black/white vertex exchange and a local exchange of edge-colors 0 and 1 in the second TS-configuration:

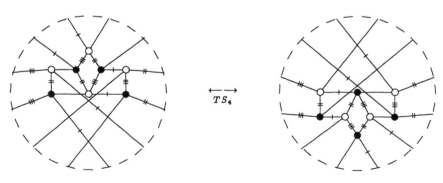

The fifth TS-move is a local π-rotation of the third TS-configuration:

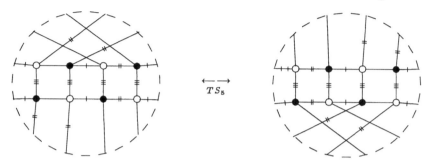

The sixth TS-moves is a local π-rotation of the fourth TS-configuration:

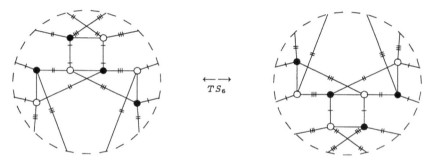

Theorem 14 *Two 3-gems which differ by a TS-move represent the same 3-manifold.*

Proof: The result folllows because each TS-move is factorizable as a finite sequence of moves where each move is either a 1-dipole move or a 2-dipole move [LD91]. ∎

We say that two rigid 3-gems are in the same $TS\rho$-*class* if they are linked by a finite number of TS- and ρ-moves to another 3-gem. The number of vertices of this third 3-gem is at most the minimum of the first two. Therefore it is a conceptually easy problem to decide whether two 3-gems are in the same $TS\rho$-class. By definition, they are in the same TS-*class* if they are in the same $TS\rho$-class and have the same number of vertices. A TS-class is *essential* if all its constituent 3-gems are rigid. The computation of the essential TS-classes was central in the topological classification of the 9351 rigid 3-gems up to 28 vertices [LD91]; this was done through the algorithm called $TS\rho$ based on the ρ- and TS-moves discussed above. Indeed $TS\rho$ is a well determined function from the set of 3-gems into the set of essential TS-classes made of minimal 3-gems. One characteristic of TS-class is that it can be recovered from any of its members.

In general two distinct TS-classes might correspond to the same homeomorphism class of 3-manifolds. Ad-hoc extensions were used in [LD91] to decide the 11 uncertainties that appear up to 28 vertices. The quantum invariants here treated or Milnor's Theorem on unique decomposition of 3-manifodls can be used to decide for the single uncertainty that is left in [LD91].

The six TS-moves are particular cases of an infinite family of (non-local) moves which awaits an appropriate implementation (which is entirely feasible). Once this is done the theory of 3-gem simplification will be substancially stronger than today.

Chapter 14

Tables of Quantum Invariants

14.1 Overview on the Tables

This Chapter is a long Appendix where we present many numerical tables of the Witten-Reshetikhin-Turaev invariants evaluated at $A = e^{i\pi/2r}$. This invariant at level r is named the $r^{\underline{th}}$ *quantum invariant*. Our implementation was made in C^{++}. It was based solely on the formulas of recoupling theory summarized in Chapter 9, on the shadow world theory developed in Chapter 11 and on Chapter 12, which provides a partition function to compute the invariants directly from a blackboard framed link. The computations have a non-empty intersection with those of [FG91] and [Nei92]. However obtained by diferent methods our results completely agree with those of papers on the intersection.

In the tables we display an initial segment of quantum invariants ($r = 3$ *to a variable maximum*) for the 3-manifolds induced by some simple framings ot the knots and links up to 7 crossings. The diagrams of the links are taken from [Rol76], but we sometimes have switched the crossings to maintain consistency. If not indicated, the framing is given by the blackboard one. Note that many of the 3-manifolds are recognized by the TS-algorithm on 3-gems.

Sometimes in the tables we present a facial identification scheme for a 3-manilfod given by a blink. Many of these facial schemes are new. We can apply $TS\rho$ to face identified polyhedra because is very simple to get a 3-gem from them. For instance, here is an identification scheme on the faces of a solid cube for the euclidean 3-manifold $EUCLID_1$ (in the notation of [LS92]):

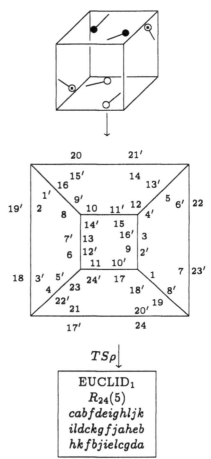

$$TS\rho\downarrow$$

$EUCLID_1$
$R_{24}(5)$
$cabfdeighljk$
$ildckgfjaheb$
$hkfbjielcgda$

The way to prove this fact is via a 3-gem inducing the same 3-manifold: consider the graph whose vertices are the sequences (v, e, f), i.e., a vertex incident to an edge incident to a face of the boundary of the cube. For $i \in \{0, 1, 2\}$ link (v, e, f) to (v', e', f') if these triples differ exactly on the $(i+1)$-coordinate. Draw an edge of color 3 between these triples if they become identified in the facial scheme. The resulting $(3+1)$-graph is a 3-gem inducing the same 3-manifold as the one given by the facial identification. An appropriate labeling of the triples is depicted in the second picture, which is ready to be fed to $TS\rho$. Note that the ends of a 3-colored edges are labeled i and i'. $TS\rho$ produces a TS-class with a single 3-gem, $R_{24}(5)$, the canonical 3-gem for $EUCLID_1$, whose code is indicated.

We identified a few simple blinks which are not present in the tables and which merit to be mentioned, as , which induces a 3-manifold apparently having the same quantum invariants as the 3-torus, induced by

. From the six euclidean 3-manifolds, $EUCLID_1$ and $EUCLID_3$,[1] are not present in the tables. We yet do not have a simple explicit blink inducing $EUCLID_1$, however, induces $EUCLID_3$:

| $R_{24}(6) \in C_{24}(39)$ |
| TS-class with |
| a single 3-gem |
| $EUCLID_3$, [LS92] |

TS_ρ applied to a 3-gem inducing the 3-manifold obtained from a solid cube by identifying by translations two of their opposite faces and rotating one face of the third pair by π before identifying with its opposite produces the $39\underline{th}$ TS-class with 24 vertices, $C_{24}(39)$. This TS-class is formed by the single rigid 3-gem $R_{24}(6)$.

The quantum invariants for the 3-manifold $EUCLID_3$ seem to always agree with the ones for the 3-torus. Specifically, with the $S^1 \times S^2$-normalization, its value at r is the integer $r - 1$ up to 8 decimal places and up to $r = 20$, for which we have computed. The homology of the manifold induced by $R_{24}(6)$ is $Z_2 \oplus Z_2 \oplus Z$, hence $EUCLID_3$ is not homeomorphic

---

[1] This notation is taken from [LS92]

to the 3-torus.

Another pair of non-homeomorphic 3-manifolds which keep having the same quantum invariants are the ones induced by the blinks:

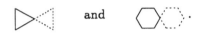

At the level r the common quantum invariant seems to be the integer $\lfloor(r+1)/3\rfloor$. The manifolds induced by these blinks are distinct because the first has homology Z and the second has homology $Z_3 \oplus Z_3 \oplus Z$.

The appearance of these integer valued quantum invariants is a point in favor of the $S^1 \times S^1$-normalization, as compared with the S^3-normalization. In displaying the tables we indicate the computation time taken in a 486-machine running at 33 MHertz and the number of admissible "r-states". An r-state is a q-admissible state with $q = e^{i\pi/r}$. An empty row indicates that the invariant vanishes for that value of r.

If on the reach of our program, we also provide few terms of the derived series of the fundamental groups. Thus G' means the derived subgroup, the subgroup of G generated by the commutators. These derived subgroups are obtained from 3-gems by a covering technique discussed in [LS92].

The *TS-code of a 3-gem* is the lexicographically minimum code among the codes of the 3-gems belonging to its image under $TS\rho$.

We give the TS-code of the 3-manifolds appearing in the tables. Observe that they are completely classified: whenever the quantum invariants seem to agree up to conjugation (up to the precision used) they have the same TS-code. *And having the same TS-code is a definite proof that they are indeed homeomorphic up to orientation.*

In the case of the manifolds $EUCLID_3$ and $S^1 \times S^1 \times S^1$ discussed on the previous two pages, we have seen, we seen computationally that they give the same invariants for r less that or equal to twenty. In fact it is easy to prove that these manifolds have the same invariants for all r. The fist step in the proof is to transform the link we used for $EUCLID_3$ by a handle slide as shown below.

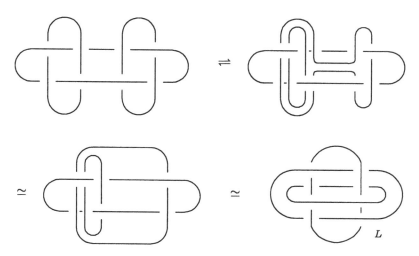

Thus $EUCLID_3$ is given by surgery in the blackboard framing on a link L that is obtained from the Borromean rings by switching two crossings. We will now show that this surgery has the same invariants as the Borromean rings. First recall by Lemma 22 the following basic result on the unnormalized invariant (due to Lickorish [Lic93]): (Diagrammatic equations apply to the evaluation of the unnormalized bracket calculation obtained by putting an ω on each link component.)

$$\left\langle \underset{\omega}{\overset{i}{}} \right\rangle = 0 \quad \text{if} \quad i \neq 0. \qquad \left[\left(\right)^{\omega} = \sum_i \Delta_i \left(\right)_i \right].$$

Apply this result to one of the rings in the link L as shown below.

Borromean.

We conclude that L and the Borromean rings have the same unnormalized invariants for all r. It is also easy to check that while the linking matrices for these two links implicate different homology groups for EUCLID$_3$ and $S^1 \times S^1 \times S^1$, the signatures and nullities are unchanged. Hence the two manifolds have the same Witten-Reshetikhin-Turaev invariants for all values of the root of unity. This appears to be a new type of coincidence of invariants. It does not seem to derive from properties of mutation as previous examples of this sort of phenomemon. In any case, the link L and

Borromean rings are certainly not mutations of each other, as a check of their Jones polynomials reveals.

This example raises many questions. We plan to investigate these a separate publication. We thank Hugh Morton for helpful conversation in relation to this example and particularly for suggesting to us the Lickorish trick.

The same manifolds have been independently checked to have the same Witten $SU(2)$ invariants by Rong [Ron93] using gluing techniques in topological quantum field theory.

14.2 Knot 3_1

Table 1: Values of the quantum invariants for the 3-manifold induced by $3_1(-3)$
(trefoil with framing number -3 – six negative curls)

a single rigid 3-gem
with 32 vertices.
$\pi_1/\pi_1' = Z_3$
$\pi_1'/\pi_1'' = Z_2 \oplus Z_2$
$\pi_1''/\pi_1''' = Z_3 \oplus Z_3 \oplus Z_6$

r	$S^1 \times S^2$ – normalization	S^3 – normalization	C^{++}, 486 time	$r-$ states
3	$-0.70710678i$	$-1.00000000i$	$0.007sec$	2
4	$+0.35355339 + 0.35355339i$	$+0.70710678 + 0.70710678i$	$0.009sec$	4
5	$+0.57206140 + 0.18587402i$	$+1.53884177 + 0.50000000i$	$0.014sec$	6
6	$+0.50000000i$	$+1.73205081i$	$0.021sec$	9
7	$+0.04478304 + 0.61411384i$	$+0.19309643 + 2.64794847i$	$0.028sec$	12
8	$-0.17677670 - 0.42677670i$	$-0.92387953 - 2.23044250i$	$0.037sec$	16
9	$-1.01801641 - 0.58775205i$	$-6.31406943 - 3.64542968i$	$0.046sec$	20
10	$-0.29270510 + 0.21266270i$	$-2.11803399 + 1.53884177i$	$0.058sec$	25
11	$+0.97018318 + 0.01799223i$	$+8.07603234 + 0.14977151i$	$0.069sec$	30
12	$+0.70710678 - 0.35355339i$	$+6.69213043 - 3.34606521i$	$0.082sec$	36
13	$-0.09318293 + 0.28964590i$	$-0.99270887 + 3.08569459i$	$0.096sec$	42
14	$-0.21231962 + 0.43023302i$	$-2.52445867 + 5.11542687i$	$0.112sec$	49
15	$-0.65123069 - 0.14548807i$	$-8.57801059 - 1.91636888i$	$0.128sec$	56
16	$-0.70053205 - 0.00130780i$	$-10.15634101 - 0.01896048i$	$0.146sec$	64
17	$+0.50829065 + 0.11012563i$	$+8.06483292 + 1.74731682i$	$0.165sec$	72
18	$+1.10286853 - 0.63674144i$	$+19.05350025 - 11.00054350i$	$0.185sec$	81
19	$+0.02541632 - 0.47564637i$	$+0.47594739 - 8.90697915i$	$0.206sec$	90
20	$-0.55902104 + 0.72539254i$	$-11.30044943 + 14.66360131i$	$0.229sec$	100

TS-code:

dabchefgjilknmpo
jpfedcbkhniolamg
glimnjadpkfbhoec

Table 2: Values of the quantum invariants for the 3-manifold induced by $3_1(-2)$
(trefoil with framing number -2 – five negative curls)

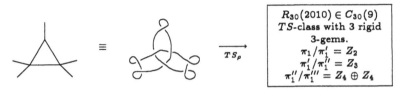

$$R_{30}(2010) \in C_{30}(9)$$
TS-class with 3 rigid
3-gems.
$$\pi_1/\pi_1' = Z_2$$
$$\pi_1'/\pi_1'' = Z_3$$
$$\pi_1''/\pi_1''' = Z_4 \oplus Z_4$$

r	$S^1 \times S^2$ – normalization	S^3 – normalization	C^{++}, 486 time	$r-$ states
3			0.007sec	2
4	$-0.92387953i$	$-1.84775907i$	0.008sec	4
5			0.013sec	6
6	$-0.28867513 - 0.86602540i$	$-1.00000000 - 3.00000000i$	0.019sec	9
7			0.025sec	12
8	$-0.67063260 - 0.67063260i$	$-3.50489485 - 3.50489485i$	0.033sec	16
9			0.042sec	20
10	$-0.72028931 - 0.53594128i$	$-5.21206244 - 3.87810750i$	0.052sec	25
11			0.062sec	30
12	$-0.89239910 - 0.33329334i$	$-8.44575577 - 3.15432203i$	0.074sec	36
13			0.087sec	42
14	$-0.87378336 - 0.19808219i$	$-10.38919542 - 2.35517711i$	0.101sec	49
15			0.116sec	56
16	$-0.95089370 - 0.01459786i$	$-13.78609414 - 0.21164035i$	0.132sec	64
17			0.149sec	72
18	$-0.89134384 + 0.11356637i$	$-15.39913386 + 1.96200795i$	0.167sec	81
19			0.186sec	90
20	$-0.90657823 + 0.27437797i$	$-18.32621780 + 5.54647166i$	0.206sec	100
21			0.227sec	110
22	$-0.81510340 + 0.38824089i$	$-18.99585574 + 9.04789255i$	0.250sec	121
23			0.272sec	132

TS-code:
$dabcgefjhimklon$
$jomedlhgknbfcai$
$lfhnibaoekjgdmc$

Table 3: Values of the quantum invariants for the 3-manifold induced by $3_1(-1)$
(trefoil with framing number -1 – four negative curls)

$R_{28}(172) \in C_{28}(3)$
TS-class made of
6 rigid 3-gems
homology sphere
$\pi_1 = \pi' = \langle 7,3,2 \rangle$

r	$S^1 \times S^2$ – *normalization*	S^3 – *normalization*	$C^{++},486$ *time*	$r-$ *states*
3	$+0.70710678$	$+1.00000000$	$0.006sec$	2
4	-0.50000000	-1.00000000	$0.008sec$	4
5	$-0.71637742 - 0.35355339i$	$-1.92705098 - 0.95105652i$	$0.011sec$	6
6	$+0.28867513$	$+1.00000000$	$0.017sec$	9
7	$+0.58547400 + 0.73416118i$	$+2.52445867 + 3.16557105i$	$0.022sec$	12
8	$+0.19134172i$	$+1.00000000i$	$0.030sec$	16
9	$+0.09888884 - 0.87838728i$	$+0.61334080 - 5.44804403i$	$0.037sec$	20
10	$+0.38819660 - 0.50655533i$	$+2.80901699 - 3.66546879i$	$0.046sec$	25
11	$-0.36522184 + 0.40677809i$	$-3.04019223 + 3.38611619i$	$0.056sec$	30
12	$-0.78867513 + 0.18301270i$	$-7.46410162 + 1.73205081i$	$0.067sec$	36
13	$+0.24724342 - 0.10586787i$	$+2.63396677 - 1.12784583i$	$0.078sec$	42
14	$+0.85966457 + 0.41399264i$	$+10.22132448 + 4.92233044i$	$0.090sec$	49
15	$+0.08219527 + 0.20412415i$	$+1.08267606 + 2.68872324i$	$0.103sec$	56
16	$-0.26138010 - 0.81695034i$	$-3.78949898 - 11.84417797i$	$0.118sec$	64
17	$+0.16713092 - 0.59182253i$	$+2.65179560 - 9.39019802i$	$0.133sec$	72
18	$-0.19211727 + 0.53326852i$	$-3.31907786 + 9.21291311i$	$0.149sec$	81
19	$-0.67439974 + 0.46666858i$	$-12.62884527 + 8.73886052i$	$0.166sec$	90
20	$+0.19737596 - 0.16561496i$	$+3.98989828 - 3.34785876i$	$0.184sec$	100

The fundamental group of this homology sphere induced by $R_{28}(172)$ is $\langle 7,3,2 \rangle = \langle a,b,c \mid a^7 = b^3 = c^2 = abc \rangle$. See [LD89], [LD91], [LS92]. Other blinks inducing the same oriented 3-manifold:

TS-code:

$dabcgefjhilknm$
$jlmedchgnaibfk$
$nfikmjaecbdhlg$

Table 4: Values of the quantum invariants for the 3-manifold induced by $3_1(0)$ (trefoil with framing number 0 – three negative curls)

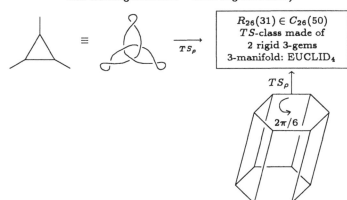

r	$S^1 \times S^2$ – normalization	S^3 – normalization	$C^{++}, 486$ time	$r-$ states
3	+1.00000000	+1.41421356	0.006sec	2
4			0.007sec	4
5	−0.30901699 − 0.95105652i	−0.83125388 − 2.55833637i	0.010sec	6
6	+0.50000000 − 0.86602540i	+1.73205081 − 3.00000000i	0.015sec	9
7			0.020sec	12
8	−0.70710678 − 0.70710678i	−3.69551813 − 3.69551813i	0.026sec	16
9	+0.17364818 − 0.98480775i	+1.07702256 − 6.10809849i	0.033sec	20
10			0.041sec	25
11	−0.84125353 − 0.54064082i	−7.00279171 − 4.50042096i	0.049sec	30
12	−1.00000000i	−9.46410162i	0.059sec	36
13			0.069sec	42
14	−0.90096887 − 0.43388374i	−10.71242836 − 5.15883360i	0.080sec	49
15	−0.10452846 − 0.99452190i	−1.37684892 − 13.09984239i	0.092sec	56
16			0.105sec	64
17	−0.93247223 − 0.36124167i	−14.79514281 − 5.73166886i	0.118sec	72
18	−0.17364818 − 0.98480775i	−3.00000000 − 17.01384546i	0.132sec	81

This manifold can be obtained from a solid hexagonal prism by identifying via translations the three pairs of opposite lateral faces and the bases by a $\pi/3$-rotation in the upper one followed by a translation. It is the euclidean 3-manifold whose fundamental group (see EUCLID$_4$ in [LS92]) is

$$\langle a, b, c \mid a^6 = b^3 = c^2 = abc \rangle.$$

Another blink inducing the same manifold: 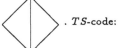 . TS-code:

$dabcgefihkjml$
$ijmedckgalhbf$
$lfhkjbaceimgd$

Table 5: Values of the quantum invariants for the 3-manifold induced by $3_1(1)$ (trefoil with framing number 1 – two negative curls)

			$R_{24}(2) \in C_{24}(1)$
		$\xrightarrow{TS_\rho}$	TS-class with a
			single 3-gem
			Poincaré's sphere: $S^3/\langle 5,3,2 \rangle$

r	$S^1 \times S^2$ – normalization	S^3 – normalization	$C^{++}, 486$ time	$r-$ states
3	$+0.70710678$	$+1.00000000$	$0.006 sec$	2
4	-0.50000000	-1.00000000	$0.006 sec$	4
5	$-0.30075048 + 0.92561479i$	$-0.80901699 + 2.48989828i$	$0.009 sec$	6
6	$+0.28867513$	$+1.00000000$	$0.014 sec$	9
7	$-0.84603445 + 0.04478304i$	$-3.64794847 + 0.19309643i$	$0.018 sec$	12
8	$+0.73253782i$	$+3.82842712i$	$0.024 sec$	16
9	$-0.17612688 - 0.40204608i$	$-1.09239627 - 2.49362077i$	$0.030 sec$	20
10	$-0.76631190 + 0.55675818i$	$-5.54508497 + 4.02874005i$	$0.038 sec$	25
11	$+0.29986112 + 0.15573689i$	$+2.49611427 + 1.29639039i$	$0.047 sec$	30
12	$-0.78867513 - 0.18301270i$	$-7.46410162 - 1.73205081i$	$0.056 sec$	36
13	$-0.11486097 + 0.74265244i$	$-1.22365232 + 7.91172464i$	$0.064 sec$	42
14	$-0.10744239 - 0.39775226i$	$-1.27747907 - 4.72923401i$	$0.074 sec$	49
15	$-0.77709557 + 0.53440395i$	$-10.23590284 + 7.03916882i$	$0.085 sec$	56
16	$+0.31417116 + 0.17622148i$	$+4.55486585 + 2.55486585i$	$0.097 sec$	64
17	$-0.78042634 - 0.14285305i$	$-12.38269491 - 2.26658898i$	$0.109 sec$	72
18	$-0.05909505 + 0.76366977i$	$-1.02094453 + 13.19339680i$	$0.123 sec$	81
19	$-0.13018472 - 0.37301190i$	$-2.43784591 - 6.98504059i$	$0.137 sec$	90
20	$-0.70858278 + 0.62543139i$	$-14.32379682 + 12.64291553i$	$0.152 sec$	100

The manifold of $R_{24}(2)$ has fundamental group the binary icosahedral group, $\langle 5,3,2 \rangle = \langle a,b,c \mid a^5 = b^3 = c^2 = abc \rangle$. It is homeomorphic to the quotient $S^3/\langle 5,3,2 \rangle$. This is Poincaré's original homology sphere. See [LD89], [LD91], [LS92]. Other blink inducing the same oriented 3-manifold: .....

TS-code:
$dabcgefjhilknm$
$jlmedchgnaibfk$
$nfikmjaecbdhlg$

Table 6: Values of the quantum invariants for the 3-manifold induced by $3_1(2)$ (trefoil with framing number 2 — one negative curl)

$R_{24}(28) \in C_{24}(13)$
TS-class with a
single 3-gem
3-manifold: $S^3/\langle 4,3,2 \rangle$

r	$S^1 \times S^2$ — normalization	S^3 — normalization	C^{++}, 486 time	r— states
3			0.005sec	2
4	+0.92387953i	+1.84775907i	0.005sec	4
5			0.007sec	6
6	−0.28867513 − 0.86602540i	−1.00000000 − 3.00000000i	0.011sec	9
7			0.014sec	12
8	+0.31015268 + 0.86572292i	+1.62093605 + 4.52448601i	0.019sec	16
9			0.024sec	20
10	−0.23229208 − 0.90311247i	−1.68088129 − 6.53498326i	0.030sec	25
11			0.037sec	30
12	+0.09904576 + 0.94205477i	+0.93737914 + 8.91570205i	0.044sec	36
13			0.050sec	42
14	+0.06946844 − 0.95931590i	+0.82597275 − 11.40616859i	0.058sec	49
15			0.067sec	56
16	−0.25780158 + 0.94054485i	−3.73761739 + 13.63605600i	0.076sec	64
17			0.086sec	72
18	+0.45110530 − 0.87781652i	+7.79343563 − 15.16543162i	0.096sec	81
19			0.107sec	90
20	−0.63463263 + 0.76853006i	−12.82891579 + 15.53561380i	0.118sec	100

The manifold of $R_{24}(28)$ has fundamental group the binary octahedral group, $\langle 4,3,2 \rangle = \langle a,b,c \mid a^4 = b^3 = c^2 = abc \rangle$. It is homeomorphic to the quotient $S^3/\langle 4,3,2 \rangle$. See [LD89], [LD91], [LS92]. Other blinks inducing the same oriented 3-manifold:

	$dabcgefjhilk$
TS-code:	$jiledchgkabf$
	$kfhjibacelgd$

Table 7: Values of the quantum invariants for the 3-manifold by 3(3) (trefoil with framing number 3 – the simplest projection)

$R_{22}(1) \in C_{22}(1)$
TS-class with a
single 3-gem
3-manifold: $S^3/\langle 3, 3, 2\rangle$

r	$S^1 \times S^2$ − normalization	S^3 − normalization	$C^{++}, 486$ time	$r-$ states
3	$+0.70710678i$	$+1.00000000i$	$0.004sec$	2
4	$+0.35355339 - 0.35355339i$	$+0.70710678 - 0.70710678i$	$0.004sec$	4
5	$-0.21850801 + 0.90225143i$	$-0.58778525 + 2.42705098i$	$0.006sec$	6
6	$-0.50000000i$	$-1.73205081i$	$0.009sec$	9
7	$-0.50805530 + 0.53386681i$	$-2.19064313 + 2.30193774i$	$0.012sec$	12
8	$+0.32322330 - 0.78033009i$	$+1.68924640 - 4.07820156i$	$0.015sec$	16
9	$-0.35355339 + 0.34590729i$	$-2.19285330 + 2.14542968i$	$0.019sec$	20
10	$+0.76631190 - 0.39429833i$	$+5.54508497 - 2.85316955i$	$0.024sec$	25
11	$-0.56088532 + 0.45576490i$	$-4.66894094 + 3.79389393i$	$0.029sec$	30
12	$+0.61237244$	$+5.79555496$	$0.034sec$	36
13	$-0.88793114 + 0.01176234i$	$-9.45942723 + 0.12530814i$	$0.046sec$	42
14	$+0.60899921 + 0.02025142i$	$+7.24093881 + 0.24078731i$	$0.054sec$	49
15	$-0.58835672 - 0.45271780i$	$-7.74983465 - 5.96319882i$	$0.054sec$	56
16	$+0.71509707 + 0.42546890i$	$+10.36750525 + 6.16846475i$	$0.061sec$	64
17	$-0.34952089 - 0.44947463i$	$-5.54570031 - 7.13162409i$	$0.069sec$	72
18	$+0.25000000 + 0.78712518i$	$+4.31907786 + 13.59861971i$	$0.077sec$	81
19	$-0.27176891 - 0.67277595i$	$-5.08915893 - 12.59843795i$	$0.086sec$	90
20	$-0.13110234 + 0.60923960i$	$-2.65019609 + 12.31560305i$	$0.096sec$	100

The manifold of $R_{22}(1)$ has fundamental group the binary tetrahedral group, $\langle 3, 3, 2\rangle = \langle a, b, c \mid a^3 = b^3 = c^2 = abc\rangle$. It is homeomorphic to the quotient $S^3/\langle 3, 3, 2\rangle$. See [LD89], [LD91], [LS92]. Other blink inducing the same oriented 3-manifold:

TS-code:

$cabfdeighkj$
$ikdcjgfeahb$
$jehkiadcbfg$

Table 8: Values of the quantum invariants for the 3-manifold induced by $3_1(4)$ (trefoil with framing number 4 – one positive curl)

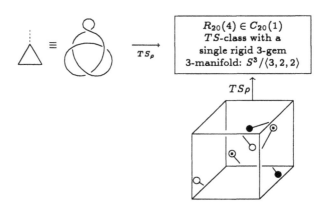

$$R_{20}(4) \in C_{20}(1)$$
TS-class with a
single rigid 3-gem
3-manifold: $S^3/\langle 3,2,2\rangle$

r	$S^1 \times S^2$ – normalization	S^3 – normalization	$C^{++}, 486$ time	$r-$ states
3	$+0.70710678 + 0.70710678i$	$+1.00000000 + 1.00000000i$	$0.005sec$	2
4	$-0.65328148 + 0.27059805i$	$-1.30656296 + 0.54119610i$	$0.005sec$	4
5	$-0.01443935 - 1.12968345i$	$-0.03884177 - 3.03884177i$	$0.007sec$	6
6	$+1.15470054$	$+4.00000000$	$0.011sec$	9
7	$-0.40843813 + 0.76906498i$	$-1.76111181 + 3.31606994i$	$0.014sec$	12
8	$-0.55368450 - 0.97130504i$	$-2.89369463 - 5.07628477i$	$0.019sec$	16
9	$+1.07791490 - 0.37080812i$	$+6.68557934 - 2.29987274i$	$0.024sec$	20
10	$-0.17082039 + 0.85065081i$	$-1.23606798 + 6.15536707i$	$0.030sec$	25
11	$-0.70274421 - 0.78095730i$	$-5.84980763 - 6.50087175i$	$0.037sec$	30
12	$+0.96574120 - 0.48374643i$	$+9.13987282 - 4.57822536i$	$0.044sec$	36
13	$-0.08661209 + 0.84632810i$	$-0.92270758 + 9.01621614i$	$0.050sec$	42
14	$-0.74851543 - 0.70449548i$	$-8.89977241 - 8.37637968i$	$0.058sec$	49
15	$+0.93234396 - 0.52296927i$	$+12.28083457 - 6.88855126i$	$0.067sec$	56
16	$-0.04958464 + 0.87691959i$	$-0.71888002 + 12.71361459i$	$0.076sec$	64
17	$-0.80735362 - 0.66456400i$	$-12.80993866 - 10.54435615i$	$0.086sec$	72
18	$+0.91959016 - 0.57735027i$	$+15.88712603 - 9.97448307i$	$0.096sec$	81

The manifold of $R_{20}(4)$ is homeomorphic to the quotient of S^3 by $\langle 3,2,2\rangle = \langle a,b,c \mid a^3 = b^2 = c^2 = abc\rangle$. It can also be obtained from a solid cube by the identification depicted on the right figure above. [LD89], [LD91], [LS92]. Other blinks inducing the same oriented

3-manifold:

TS-code:
$cabfdehgji$
$hjdcbifage$
$ifgahbjedc$

Table 9: Values of the quantum invariants for the 3-manifold induced by $3_1(5)$ (trefoil with framing number 5 – two positive curls)

			$R_{20}(2) \in C_{20}(3)$
			TS-class with a
△	≡	$\xrightarrow{TS_\rho}$	single 3-gem.
			3-manifold: $L_{5,1}$

r	$S^1 \times S^2$ – normalization	S^3 – normalization	$C^{++}, 486$ time	$r-$ states
3	$+0.70710678$	$+1.00000000$	$0.005 sec$	2
4	$+0.50000000i$	$+1.00000000i$	$0.006 sec$	4
5	$-0.41562694 + 0.57206140i$	$-1.11803399 + 1.53884177i$	$0.009 sec$	6
6	-0.28867513	-1.00000000	$0.013 sec$	9
7	$+0.11596031 - 0.50805530i$	$+0.50000000 - 2.19064313i$	$0.017 sec$	12
8	$+0.32664074 - 0.32664074i$	$+1.70710678 - 1.70710678i$	$0.023 sec$	16
9	$+0.15150649 + 0.26241694i$	$+0.93969262 + 1.62759536i$	$0.028 sec$	20
10	$-0.15450850 + 0.47552826i$	$-1.11803399 + 3.44095480i$	$0.035 sec$	25
11	$-0.22119193 + 0.06494781i$	$-1.84125353 + 0.54064082i$	$0.044 sec$	30
12	$-0.39433757i$	$-3.73205081i$	$0.052 sec$	36
13	$+0.20833408 - 0.30182402i$	$+2.21945257 - 3.21543215i$	$0.059 sec$	42
14	$+0.14692973 + 0.18424406i$	$+1.74697960 + 2.19064313i$	$0.069 sec$	49
15	$-0.08487959 + 0.39932709i$	$-1.11803399 + 5.25993637i$	$0.079 sec$	56
16	$-0.18147187 + 0.07516811i$	$-2.63098631 + 1.08979021i$	$0.090 sec$	64
17	$-0.03043965 - 0.32849616i$	$-0.48297310 - 5.21210981i$	$0.102 sec$	72
18	$+0.15661544 - 0.27126589i$	$+2.70573706 - 4.68647407i$	$0.114 sec$	81
19	$+0.13496658 + 0.14661279i$	$+2.52739140 + 2.74547893i$	$0.127 sec$	90
20	$-0.05530794 + 0.34920056i$	$-1.11803399 + 7.05898879i$	$0.141 sec$	100

The manifold induced by $R_{20}(2)$ is the lens space $L_{5,1}$. Another blink inducing the same oriented 3-manifold:

TS-code:

```
cabfdehgji
hjdcbifage
gfeicbjdha
```

Table 10: Values of the quantum invariants for the 3-manifold induced by $3_1(6)$ (trefoil with framing number 6) – three positive curls)

			$R_{18}(4) \in C_{18}(2)$ TS-class with a single 3-gem. 3-manifold: $L_{2,1} \# L_{3,1}$

r	$S^1 \times S^2$ – normalization	S^3 – normalization	C^{++}, 486 time	$r-$ states
3			0.006sec	2
4	$+0.27059805 + 0.27059805i$	$+0.54119610 + 0.54119610i$	0.007sec	4
5			0.010sec	6
6	$+0.36602540i$	$+1.26794919i$	0.015sec	9
7			0.020sec	12
8	$-0.30768885 + 0.12744889i$	$-1.60805943 + 0.66608002i$	0.026sec	16
9			0.033sec	20
10	$-0.20955371 - 0.15224969i$	$-1.51634492 - 1.10168907i$	0.041sec	25
11			0.049sec	30
12	$-0.25215724i$	$-2.38644174i$	0.059sec	36
13			0.069sec	42
14	$+0.21832071 - 0.10513771i$	$+2.59581102 - 1.25007670i$	0.080sec	49
15			0.091sec	56
16	$+0.17367171 + 0.11604372i$	$+2.51789919 + 1.68240645i$	0.105sec	64
17			0.118sec	72
18	$+0.20490387i$	$+3.53998301i$	0.132sec	81
19			0.147sec	90
20	$-0.17806871 + 0.09073054i$	$-3.59960778 + 1.83409177i$	0.163sec	100
21			0.180sec	110
22	$-0.15126916 - 0.09721479i$	$-3.52530393 - 2.26557527i$	0.198sec	121
23			0.216sec	132
24	$-0.17715600i$	$-4.70163402i$	0.235sec	144
25			0.256sec	156

TS-code:

$dabcgefih$
$ihfedcbga$
$higfbdcae$

Table 11: Values of the quantum invariants for the 3-manifold induced by $3_1(7)$ (trefoil with framing number 7 – with four positive curls)

r	$S^1 \times S^2$ – normalization	S^3 – normalization	$C^{++}, 486$ time	$r-$ states
3	$+0.70710678i$	$+1.00000000i$	$0.006sec$	2
4	$-0.35355339 - 0.35355339i$	$-0.70710678 - 0.70710678i$	$0.008sec$	4
5	$+0.35355339 - 0.48662449i$	$+0.95105652 - 1.30901699i$	$0.011sec$	6
6	$+0.28867513$	$+1.00000000$	$0.017sec$	9
7			$0.023sec$	12
8	$-0.17677670 + 0.07322330i$	$-0.92387953 + 0.38268343i$	$0.030sec$	16
9	$-0.40204608 - 0.23212141i$	$-2.49362077 - 1.43969262i$	$0.037sec$	20
10	$+0.11180340 - 0.34409548i$	$+0.80901699 - 2.48989828i$	$0.046sec$	25
11	$+0.17422296 + 0.27109622i$	$+1.45027281 + 2.25666855i$	$0.056sec$	30
12	$-0.27883877 + 0.27883877i$	$-2.63895843 + 2.63895843i$	$0.067sec$	36
13	$-0.18095040 - 0.02197136i$	$-1.92772512 - 0.23406821i$	$0.078sec$	42
14			$0.090sec$	49
15	$+0.14125017 - 0.04589496i$	$+1.86054730 - 0.60452846i$	$0.103sec$	56
16	$+0.28832037 + 0.19264951i$	$+4.18008002 + 2.79304017i$	$0.118sec$	64
17	$-0.09887822 + 0.25523411i$	$-1.56885894 + 4.04969173i$	$0.133sec$	72
18	$-0.12767407 - 0.22113798i$	$-2.20573706 - 3.82044866i$	$0.149sec$	81
19	$+0.23139619 - 0.21301520i$	$+4.33313728 - 3.98893387i$	$0.166sec$	90
20	$+0.14179689 + 0.02245842i$	$+2.86638321 + 0.45399050i$	$0.184sec$	100
21			$0.203sec$	110
22	$-0.12017874 + 0.03528766i$	$-2.80074651 + 0.82237337i$	$0.223sec$	121
23	$-0.23587240 - 0.16649692i$	$-5.87428880 - 4.14652583i$	$0.243sec$	132
24	$+0.08764269 - 0.21158817i$	$+2.32599430 - 5.61544699i$	$0.265sec$	144
25	$+0.10499053 + 0.19097707i$	$+2.96168522 + 5.38728542i$	$0.287sec$	156

TS-code:
$$
\begin{array}{l}
cabfdehgji \\
hjdcbifage \\
jheicbafdg
\end{array}
$$

Table 12: Values of the quantum invariants for the 3-manifold induced by $3_1(8)$ (trefoil with framing number 8 – five positive curls)

$$R_{22}(2) \in C_{22}(7)$$
TS-class with a
single 3-gem.
3-manifold: $S^3/(Z_3 \rtimes_i Z_8)$

r	$S^1 \times S^2$ – normalization	S^3 – normalization	$C^{++}, 486$ time	$r-$ states
3	$+0.70710678 + 0.70710678i$	$+1.00000000 + 1.00000000i$	$0.007sec$	2
4	$-0.38268343 + 0.92387953i$	$-0.76536686 + 1.84775907i$	$0.008sec$	4
5	$-1.12075944 + 0.68374342i$	$-3.01483624 + 1.83926573i$	$0.013sec$	6
6	-1.15470054	-4.00000000	$0.019sec$	9
7	$-0.80125141 - 0.89081749i$	$-3.45485204 - 3.84104490i$	$0.025sec$	12
8	$-0.27589938 - 1.38703985i$	$-1.44191964 - 7.24901957i$	$0.033sec$	16
9	$+0.45659672 - 1.16370351i$	$+2.83196163 - 7.21766823i$	$0.042sec$	20
10	$+1.17082039 - 0.52573111i$	$+8.47213595 - 3.80422607i$	$0.052sec$	25
11	$+1.36673518 + 0.12326067i$	$+11.37702419 + 1.02605077i$	$0.062sec$	30
12	$+0.91608553 + 0.70293715i$	$+8.66992654 + 6.65266861i$	$0.074sec$	36
13	$+0.23851182 + 1.14228866i$	$+2.54094609 + 12.16918282i$	$0.087sec$	42
14	$-0.29898978 + 1.16399264i$	$-3.55495813 + 13.83975429i$	$0.101sec$	49
15	$-0.70774332 + 0.69631236i$	$-9.32239493 + 9.17182644i$	$0.116sec$	56
16	$-0.99518473 + 0.09801714i$	$-14.42822712 + 1.42105634i$	$0.132sec$	64
17	$-0.95812101 - 0.30791397i$	$-15.20210114 - 4.88554082i$	$0.149sec$	72
18	$-0.56486424 - 0.57735027i$	$-9.75877048 - 9.97448307i$	$0.167sec$	81
19	$-0.12038497 - 0.80850144i$	$-2.25433538 - 15.14004066i$	$0.186sec$	90
20	$+0.16771251 - 0.85126901i$	$+3.39026001 - 17.20815780i$	$0.206sec$	100

Let $H \rtimes_t K$ denote the semi-direct product of the group H by the group K with K inducing an automorphism group of order t on H. When $H = Z_n$, $K = Z_{2m}$ and $t = 2$ we write the group as $Z_n \rtimes_i Z_{2m}$. The manifold induced by $R_{22}(2)$ is identified as $S^3/(Z_3 \rtimes_i Z_8)$, [LS92].

TS-code:

$cabfdeighkj$
$ikdcjgfeahb$
$kdigahbfjce$

Table 13: Values of the quantum invariants for the 3-manifold induced by $3_1(9)$ (trefoil with framing number 9 – six positive curls)

$$R_{26}(8) \in C_{26}(18)$$
TS-class with 11
rigid 3-gems.
3-manifold: $S^3/(Q_8 \rtimes_3 Z_9)$

r	$S^1 \times S^2$ – normalization	S^3 – normalization	C^{++}, 486 time	$r-$ states
3	+0.70710678	+1.00000000	0.007sec	2
4	+0.50000000	+1.00000000	0.009sec	4
5	+0.55762205 − 0.57206140i	+1.50000000 − 1.53884177i	0.014sec	6
6	−0.50000000i	−1.73205081i	0.021sec	9
7	−0.23759308 − 0.22610588i	−1.02445867 − 0.97492791i	0.028sec	12
8	−0.73253782	−3.82842712	0.037sec	16
9	−0.45451948 + 0.54167522i	−2.81907786 + 3.35964617i	0.046sec	20
10	−0.01631190 + 0.63798811i	−0.11803399 + 4.61652531i	0.058sec	25
11	+0.42528742 + 0.82454339i	+3.54019223 + 6.86369258i	0.069sec	30
12	+0.75000000 + 0.25000000i	+7.09807621 + 2.36602540i	0.082sec	36
13	+0.54755931 − 0.24768552i	+5.83333237 − 2.63867661i	0.096sec	42
14	+0.37220062 − 0.67122029i	+4.42542754 − 7.98074105i	0.112sec	49
15	−0.31622777 − 0.60150096i	−4.16535213 − 7.92297057i	0.128sec	56
16	−0.62454508 − 0.09754516i	−9.05467898 − 1.41421356i	0.147sec	64
17	−0.78334078 + 0.30632948i	−12.42893717 + 4.86040035i	0.165sec	72
18	−0.29619814 + 0.81379768i	−5.11721119 + 14.05942220i	0.185sec	81
19	+0.33228574 + 0.73207401i	+6.22240040 + 13.70885654i	0.206sec	90
20	+0.71173789 + 0.54104636i	+14.38757654 + 10.93709629i	0.229sec	100
21	+0.80016926 − 0.16756750i	+17.39670765 − 3.64313270i	0.251sec	110
22	+0.31632111 − 0.62931365i	+7.37181345 − 14.66605512i	0.277sec	121
23	−0.09314593 − 0.74612841i	−2.31975475 − 18.58196973i	0.302sec	132
24	−0.69290965 − 0.36626891i	−18.38948486 − 9.72059856i	0.329sec	144

The manifold of $R_{26}(8)$ is identified as $S^3/(Q_8 \rtimes_3 Z_9)$, [LS92].

TS-code:
cabfdeighkjml
imdcbjfkaglhe
mgkjlcbadifeh

Table 14: Values of the quantum invariants for the 3-manifold induced by $3_1(10)$
(trefoil with framing number 10 – seven positive curls)

$$R_{30}(432) \in C_{30}(42)$$
TS-class with 105 rigid
3-gems.
$$\pi/\pi' = Z_{10}$$
$$\pi'/\pi'' = Z_3 \quad (S^3/\langle 3,3,2\rangle)$$
$$\pi''/\pi''' = Z_2 \oplus Z_2 \quad (S^3/\langle 2,2,2\rangle)$$
$$\pi'''/\pi^{iv} = Z_2 \quad (L_{2,1})$$
$$\pi^{iv} = \pi^v \quad (S^3)$$

r	$S^1 \times S^2 - normalization$	$S^3 - normalization$	$C^{++}, 486$ time	$r-$ states
3			$0.008sec$	2
4	-0.38268343	-0.76536686	$0.010sec$	4
5			$0.016sec$	6
6	$+0.28867513 - 0.86602540i$	$+1.00000000 - 3.00000000i$	$0.023sec$	9
7			$0.031sec$	12
8	$+0.19509032 + 0.69351992i$	$+1.01959116 + 3.62450979i$	$0.041sec$	16
9			$0.051sec$	20
10	$-0.63003676 + 0.32101976i$	$-4.55898879 + 2.32292081i$	$0.063sec$	25
11			$0.076sec$	30
12	$+0.21892515 - 0.89239910i$	$+2.07192983 - 8.44575577i$	$0.091sec$	36
13			$0.106sec$	42
14	$+0.58874779 + 0.25128616i$	$+7.00015150 + 2.98776691i$	$0.124sec$	49
15			$0.141sec$	56
16	$-0.60317396 + 0.75593946i$	$-8.74483971 + 10.95963991i$	$0.161sec$	64
17			$0.182sec$	72
18	$-0.28023304 - 0.72467718i$	$-4.84139320 - 12.51974862i$	$0.204sec$	81
19			$0.227sec$	90
20	$+0.79578236 - 0.34165836i$	$+16.08651124 - 6.90652535i$	$0.252sec$	100
21			$0.277sec$	110
22	$-0.14316347 + 0.94892580i$	$-3.33640195 + 22.11456557i$	$0.305sec$	121

TS-code:
$$dabcgefjhimklon$$
$$jmledcogkainbfh$$
$$hlgfodcmenjbaki$$

14.3 Knot 4_1

Table 15: Values of the quantum invariants for the 3-manifold induced by $4_1(0)$

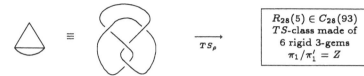

$R_{28}(5) \in C_{28}(93)$
TS-class made of
6 rigid 3-gems
$\pi_1/\pi_1' = Z$

r	$S^1 \times S^2$ – normalization	S^3 – normalization	$C^{++}, 486$ time	$r-$ states
3	+1.00000000	+1.41421356	0.005sec	2
4			0.008sec	6
5	−0.61803399	−1.66250775	0.014sec	10
6			0.027sec	19
7	+1.00000000	+4.31182025	0.042sec	28
8	+1.00000000	+5.22625186	0.065sec	44
9			0.091sec	60
10	−0.61803399	−4.47213595	0.129sec	85
11			0.171sec	110
12	+1.00000000	+9.46410162	0.226sec	146
13	+1.00000000	+10.65333422	0.285sec	182
14			0.362sec	231
15	−0.61803399	−8.14074369	0.444sec	280
16			0.546sec	344
17	+1.00000000	+15.86657740	0.653sec	408

Let M be a 2×2 matrix over the integers. Let ZM indicate the semi-direct product group

$$(Z \times Z) \rtimes_M Z = H \rtimes \langle a \rangle$$

with a acting on the canonical basis of H as the matrix M. The fundamental group of the 3-manifold of $R_{28}(5)$ is $Z_{-1\ 3}^{0\ 1}$. See [LD89], [LD91], [LS92]. Other blinks inducing the

same orientable 3-manifold:

TS-code:

dabcgefjhilknm
jlmedchgnaibfk
nfikmjaecbdhlg

Table 16: Values of the quantum invariants for the 3-manifold induced by $4_1(1)$

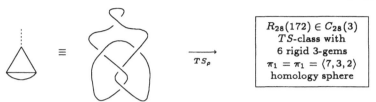

$$\triangle \quad \equiv \quad \overbrace{\qquad}^{} \quad \xrightarrow{TS_\rho}$$

$R_{28}(172) \in C_{28}(3)$
TS-class with
6 rigid 3-gems
$\pi_1 = \pi_1 = \langle 7,3,2 \rangle$
homology sphere

r	$S^1 \times S^2$ − normalization	S^3 − normalization	$C^{++},486$ time	$r-$ states
3	+0.70710678	+1.00000000	0.006sec	2
4	−0.50000000	−1.00000000	0.009sec	6
5	−0.71637742 + 0.35355339i	−1.92705098 + 0.95105652i	0.017sec	10
6	+0.28867513	+1.00000000	0.031sec	19
7	+0.58547400 − 0.73416118i	+2.52445867 − 3.16557105i	0.048sec	28
8	−0.19134172i	−1.00000000i	0.075sec	44
9	+0.09888884 + 0.87838728i	+0.61334080 + 5.44804403i	0.104sec	60
10	+0.38819660 + 0.50655533i	+2.80901699 + 3.66546879i	0.148sec	85
11	−0.36522184 − 0.40677809i	−3.04019223 − 3.38611619i	0.196sec	110
12	−0.78867513 − 0.18301270i	−7.46410162 − 1.73205081i	0.259sec	146
13	+0.24724342 + 0.10586787i	+2.63396677 + 1.12784583i	0.326sec	182
14	+0.85966457 − 0.41399264i	+10.22132448 − 4.92233044i	0.414sec	231
15	+0.08219527 − 0.20412415i	+1.08267606 − 2.68872324i	0.507sec	280
16	−0.26138010 + 0.81695034i	−3.78949898 + 11.84417797i	0.624sec	344
17	+0.16713092 + 0.59182253i	+2.65179560 + 9.39019802i	0.744sec	408
18	−0.19211727 − 0.53326852i	−3.31907786 − 9.21291311i	0.893sec	489
19	−0.67439974 − 0.46666858i	−12.62884527 − 8.73886052i	1.048sec	570
20	+0.19737596 + 0.16561496i	+3.98989828 + 3.34785876i	1.232sec	670

The fundamental group of the homology sphere induced by $R_{28}(172)$ is $\langle 7,3,2 \rangle = \langle a,b,c \mid a^7 = b^3 = c^2 = abc \rangle$. See [LD89], [LD91], [LS92]. Other blinks inducing the same oriented 3-manifold: $\times$, , . For any integer n, $M^3(4_1(n)) = -M^3(4_1(-n))$

TS-code:

$dabcgefjhilknm$
$jlmedchgnaibfk$
$nfikmjaecbdhlg$

Table 17: Values of the quantum invariants for the 3-manifold induced by $4_1(2)$

$$R_{28}(202) \in C_{28}(6)$$
TS-class with
a single rigid 3-gem.
$$\pi_1/\pi_1' = Z_2$$
$$\pi_1'/\pi_1'' = Z_5$$
$$\pi_1''/\pi_1''' = Z_2 \oplus Z_2 \oplus Z_2 \oplus Z_2$$

r	$S^1 \times S^2$ − normalization	S^3 − normalization	C^{++}, 486 time	$r-$ states
3			0.006sec	2
4	+0.92387953i	+1.84775907i	0.011sec	6
5			0.019sec	10
6	−0.78867513	−2.73205081	0.037sec	19
7			0.055sec	28
8	−0.02288733 − 0.80858229i	−0.11961494 − 4.22585467i	0.086sec	44
9			0.120sec	60
10	+0.88122428 + 0.09409548i	+6.37659882 + 0.68088129i	0.170sec	85
11			0.223sec	110
12	−0.27059805 + 0.84300901i	−2.56096744 + 7.97832291i	0.296sec	146
13			0.373sec	182
14	−0.74645259 − 0.36848811i	−8.87524547 − 4.38128627i	0.474sec	231
15			0.579sec	280
16	+0.42852736 − 0.70122874i	+6.21280645 − 10.16644168i	0.712sec	344
17			0.850sec	408
18	+0.66819110 + 0.54060056i	+11.54387757 + 9.33958364i	1.020sec	489
19			1.196sec	570
20	−0.66610546 + 0.56747261i	−13.46512996 + 11.47129529i	1.404sec	670

The fundamental group of the 3-manifold induced by $R_{28}(202)$ is $\langle 5,5,2 \rangle 2 = \langle a,b,c \mid a^5 = b^5 = (ab)^2 = c^2 \rangle$. That is, the group is an extension of $\langle 5,5,2 \rangle$ by an infinite cyclic group generated by c, which interchanges a and b and satisfies $c^2 = (ab)^2$, [LS92].

TS-code:
```
dabcgefjhilknm
jlnedchgmaibkf
mfikljaecbnhgd
```

Table 18: Values of the quantum invariants for the 3-manifold induced by $4_1(3)$

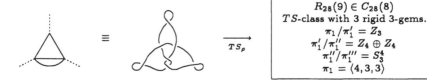

$$R_{28}(9) \in C_{28}(8)$$
TS-class with 3 rigid 3-gems.
$$\pi_1/\pi_1' = Z_3$$
$$\pi_1'/\pi_1'' = Z_4 \oplus Z_4$$
$$\pi_1''/\pi_1''' = S_3^4$$
$$\pi_1 = \langle 4, 3, 3 \rangle$$

r	$S^1 \times S^2 - normalization$	$S^3 - normalization$	$C^{++}, 486$ time	$r-$ states
3	$+0.70710678i$	$+1.00000000i$	$0.007sec$	2
4	$+0.35355339 - 0.35355339i$	$+0.70710678 - 0.70710678i$	$0.012sec$	6
5	$+0.35355339 + 0.48662449i$	$+0.95105652 + 1.30901699i$	$0.021sec$	10
6	$-0.50000000i$	$-1.73205081i$	$0.040sec$	19
7	$+0.91548402 + 0.02296736i$	$+3.94740253 + 0.09903113i$	$0.061sec$	28
8	$-0.03033009 + 0.07322330i$	$-0.15851267 + 0.38268343i$	$0.095sec$	44
9	$+0.66446302 - 0.38362791i$	$+4.12121613 - 2.37938524i$	$0.132sec$	60
10	$+0.11180340 + 0.34409548i$	$+0.80901699 + 2.48989828i$	$0.187sec$	85
11	$+0.04958903 - 0.73276764i$	$+0.41279068 - 6.09972967i$	$0.245sec$	110
12	$+0.61237244$	$+5.79555496$	$0.325sec$	146
13	$-0.21008981 - 0.44423541i$	$-2.23815698 - 4.73258825i$	$0.409sec$	182
14	$+0.73721382 - 0.40998160i$	$+8.76539748 - 4.87463956i$	$0.520sec$	231
15	$-0.22473227 + 0.11387792i$	$-2.96017343 + 1.50000000i$	$0.635sec$	280
16	$+0.15748644 - 0.79173781i$	$+2.28324459 - 11.47864572i$	$0.781sec$	344
17	$+0.10339902 + 0.10059154i$	$+1.64058852 + 1.59604352i$	$0.932sec$	408
18	$-0.35286853 - 0.78107901i$	$-6.09626666 - 13.49416426i$	$1.117sec$	489

The fundamental group of the 3-manifold induced by $R_{28}(9)$ is ,[LS92], $\langle 4, 3, 3 \rangle = \langle a, b, c \mid a^4 = b^3 = c^3 = abc \rangle$ Let $g, r \geq 1$. We define
$S_g^r = \langle a_1, \ldots, a_g, b_1, \ldots b_g, c \mid [c, a_i] = e = [c, b_i](1 \leq i \leq g), c^r = \prod_{i=1}^g [a_i, b_i] \rangle$,
which is the fundamental group of a Seifert manifold over a surface of of genus g [Orl72].

TS-code:

$cabfdeighljknm$
$indcmgfjlheakb$
$jeingabdkfcmlh$

Table 19: Values of the quantum invariants for the 3-manifold induced by $4_1(4)$

$$R_{30}(171) \in C_{30}(19)$$
TS-class with 34 rigid 3-gems
$$\pi_1/\pi_1' = Z_4$$
$$\pi_1'/\pi_1'' = Z_3 \oplus Z_{15}$$

r	$S^1 \times S^2 - normalization$	$S^3 - normalization$	$C^{++}, 486$ time	$r-$ states
3	$+0.70710678 + 0.70710678i$	$+1.00000000 + 1.00000000i$	$0.007sec$	2
4	$-0.65328148 + 0.27059805i$	$-1.30656296 + 0.54119610i$	$0.013sec$	6
5	$-0.90544588 - 0.67569295i$	$-2.43564402 - 1.81761003i$	$0.024sec$	10
6	$+0.28867513 - 0.50000000i$	$+1.00000000 - 1.73205081i$	$0.044sec$	19
7	$+1.11352608 + 0.66131432i$	$+4.80132432 + 2.85146849i$	$0.067sec$	28
8	$+0.36729332 + 1.09584764i$	$+1.91956738 + 5.72717578i$	$0.105sec$	44
9	$-0.87406796 + 0.10365846i$	$-5.42125421 + 0.64292356i$	$0.145sec$	60
10	$-0.86180340 - 0.95105652i$	$-6.23606798 - 6.88190960i$	$0.205sec$	85
11	$+0.38090687 - 0.57903555i$	$+3.17075810 - 4.82002771i$	$0.269sec$	110
12	$+1.12560248 + 0.73052058i$	$+10.65281626 + 6.91372100i$	$0.356sec$	146
13	$+0.32982877 + 1.17236225i$	$+3.51377613 + 12.48956684i$	$0.449sec$	182
14	$-0.92596995 + 0.10903451i$	$-11.00968868 + 1.29640921i$	$0.570sec$	231
15	$-0.90481576 - 1.02896846i$	$-11.91823313 - 13.55357255i$	$0.695sec$	280
16	$+0.39163473 - 0.67208925i$	$+5.67793563 - 9.74397625i$	$0.855sec$	344
17	$+1.18253626 + 0.72275223i$	$+18.76280316 + 11.46760419i$	$1.020sec$	408
18	$+0.34774441 + 1.24598110i$	$+6.00774079 + 21.52595759i$	$1.223sec$	489
19	$-0.98330348 + 0.14397843i$	$-18.41339295 + 2.69614776i$	$1.433sec$	570
20	$-0.95694830 - 1.09455886i$	$-19.34443433 - 22.12619212i$	$1.683sec$	670
21	$+0.41629617 - 0.74986045i$	$+9.05081358 - 16.30292961i$	$1.944sec$	770
22	$+1.23613106 + 0.73035908i$	$+28.80783882 + 17.02090285i$	$2.252sec$	891
23	$+0.34770874 + 1.30338708i$	$+8.65951910 + 32.46022919i$	$2.566sec$	1012
24	$-1.02850103 + 0.15478523i$	$-27.29591667 + 4.10792463i$	$2.930sec$	1156
25	$-0.97941144 - 1.14231013i$	$-27.62828549 - 32.22350598i$	$3.307sec$	1300

TS-code:

```
dabcgefjhimklon
jkmedchgnabiolf
feiloadmcnjghkb
```

Table 20: Values of the quantum invariants for the 3-manifold induced by $4_1(5)$

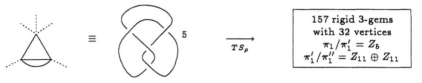

r	$S^1 \times S^2$ − *normalization*	S^3 − *normalization*	$C^{++}, 486$ *time*	$r-$ *states*
3	+0.70710678	+1.00000000	0.008sec	2
4	+0.50000000i	+1.00000000i	0.015sec	6
5	+0.25687157 + 0.35355339i	+0.69098301 + 0.95105652i	0.026sec	10
6	−0.28867513	−1.00000000	0.049sec	19
7	−0.23759308 + 0.58875155i	−1.02445867 + 2.53859088i	0.074sec	28
8	+0.05604269 − 0.05604269i	+0.29289322 − 0.29289322i	0.116sec	44
9	−0.57802870 − 0.23076559i	−3.58512231 − 1.43128334i	0.160sec	60
10	−0.05901699 − 0.18163563i	−0.42705098 − 1.31432778i	0.227sec	85
11	−0.17890674 − 0.72575072i	−1.48926169 − 6.04131920i	0.297sec	110
12	−0.18301270 − 0.07735027i	−1.73205081 − 0.73205081i	0.394sec	146
13	+0.44783261 − 0.39914726i	+4.77091205 − 4.25224921i	0.496sec	182
14	−0.05243861 − 0.52525311i	−0.62348980 − 6.24520610i	0.630sec	231
15	+0.41927136 − 0.08348209i	+5.52264232 − 1.09962613i	0.768sec	280
16	+0.40954154 − 0.46305424i	+5.93754928 − 6.71337851i	0.944sec	344
17	+0.13121136 + 0.28673378i	+2.08187525 + 4.54948373i	1.126sec	408
18	+0.66708770 + 0.23040125i	+11.52481493 + 3.98048369i	1.350sec	489
19	+0.01878787 + 0.03117295i	+0.35182260 + 0.58374623i	1.580sec	570
20	+0.18869068 + 0.54930655i	+3.81432778 + 11.10407376i	1.858sec	670
21	+0.13578693 + 0.06783508i	+2.95218236 + 1.47482169i	2.143sec	770
22	−0.39396192 + 0.56726340i	−9.18122012 + 13.21998388i	2.482sec	891
23	+0.09122886 + 0.45150261i	+2.27201085 + 11.24445571i	2.828sec	1012
24	−0.50237503 − 0.06117831i	−13.33278876 − 1.62364244i	3.232sec	1156
25	−0.38594801 + 0.31951437i	−10.88723421 + 9.01320317i	3.645sec	1300

TS-code:

```
cabfdeighljknmpo
ipjclgfendohkamb
hengmolpjicdbkfa
```

Table 21: Values of the quantum invariants for the 3-manifold induced by $4_1(6)$

			158 rigid 3-gems with 34 vertices $\pi_1/\pi_1' = Z_6$ $\pi_1'/\pi_1'' = Z_8 \oplus Z_{40}$

r	$S^1 \times S^2 - normalization$	$S^3 - normalization$	$C^{++}, 486$ time	$r-$ states
3			$0.008sec$	2
4	$+0.27059805 + 0.27059805i$	$+0.54119610 + 0.54119610i$	$0.017sec$	6
5			$0.030sec$	10
6	$+0.50000000 - 0.50000000i$	$+1.73205081 - 1.73205081i$	$0.055sec$	19
7			$0.084sec$	28
8	$-0.08515863 - 0.71538670i$	$-0.44506044 - 3.73879105i$	$0.131sec$	44
9			$0.182sec$	60
10	$-0.38122428 - 0.25655533i$	$-2.75856483 - 1.85645180i$	$0.258sec$	85
11			$0.338sec$	110
12	$-0.41741971 + 0.59246596i$	$-3.95050258 + 5.60715803i$	$0.449sec$	146
13			$0.566sec$	182
14	$+0.18543265 + 0.62456388i$	$+2.20477534 + 7.42600117i$	$0.718sec$	231
15			$0.876sec$	280
16	$+0.50555812 - 0.01786318i$	$+7.32960141 - 0.25898103i$	$1.078sec$	344
17			$1.286sec$	408
18	$+0.17760609 - 0.69613262i$	$+3.06837815 - 12.02660393i$	$1.541sec$	489
19			$1.806sec$	570
20	$-0.40549249 - 0.45353902i$	$-8.19691399 - 9.16816061i$	$2.124sec$	670
21			$2.450sec$	770
22	$-0.44681083 + 0.33174846i$	$-10.41285561 + 7.73134546i$	$2.837sec$	891
23			$3.234sec$	1012
24	$+0.12460038 + 0.69701196i$	$+3.30683333 + 18.49835823i$	$3.696sec$	1156
25			$4.169sec$	1300
26	$+0.54672233 + 0.15969175i$	$+16.35382199 + 4.77677658i$	$4.715sec$	1469

TS-code:

> $dabcgefjhimklonqp$
> $jmpednhgkaifqlcob$
> $lojfidcbmgnqephka$

14.4   Knot 5_1

Table 22: Values of the quantum invariants for the 3-manifold induced by $5_1(-1)$

r	$S^1 \times S^2$ – normalization	S^3 – normalization	$C^{++}, 486$ time	$r-$ states
3	$+0.70710678$	$+1.00000000$	$0.008sec$	2
4	-0.50000000	-1.00000000	$0.011sec$	4
5	$-0.30075048 + 0.21850801i$	$-0.80901699 + 0.58778525i$	$0.017sec$	6
6	$+0.28867513$	$+1.00000000$	$0.025sec$	9
7	$+0.74282007 + 0.40742872i$	$+3.20290660 + 1.75675939i$	$0.033sec$	12
8	$-0.73253782i$	$-3.82842712i$	$0.044sec$	16
9	$+0.61574932 - 0.54167522i$	$+3.81907786 - 3.35964617i$	$0.055sec$	20
10	$+0.04270510 + 0.13143278i$	$+0.30901699 + 0.95105652i$	$0.069sec$	25
11	$-0.53039114 + 0.61210400i$	$-4.41510021 + 5.09529727i$	$0.083sec$	30
12	-0.10566243	-1.00000000	$0.099sec$	36
13	$-0.20077610 + 0.83554512i$	$-2.13893490 + 8.90134145i$	$0.115sec$	42
14	$-0.36387160 - 0.19325679i$	$-4.32639641 - 2.29780357i$	$0.135sec$	49
15	$+0.24196815 - 0.14320045i$	$+3.18720453 - 1.88623634i$	$0.154sec$	56
16	$+0.59195628 + 0.10156364i$	$+8.58220534 + 1.47247365i$	$0.176sec$	64
17	$-0.37609061 - 0.28501124i$	$-5.96727075 - 4.52215293i$	$0.197sec$	72
18	$+0.35408086 - 0.81379768i$	$+6.11721119 - 14.05942220i$	$0.222sec$	81
19	$-0.45282842 - 0.19316291i$	$-8.47968889 - 3.61717886i$	$0.247sec$	90
20	$-0.33346023 - 0.19784718i$	$-6.74080250 - 3.99942373i$	$0.274sec$	100
21	$+0.21891407 + 0.03946755i$	$+4.75947323 + 0.85807523i$	$0.301sec$	110
22	$+0.55300163 + 1.21090466i$	$+12.88761548 + 28.21994155i$	$0.331sec$	121

TS-code:

> $fabcdelghijknmporqtsvu$
> $nspihgfedotmlruckavqjb$
> $kjimtopsnbafqclvdhrueg$

Table 23: Values of the quantum invariants for the 3-manifold induced by $5_1(0)$

r	$S^1 \times S^2 - normalization$	$S^3 - normalization$	$C^{++}, 486$ time	$r-$ states
3	$+1.00000000$	$+1.41421356$	$0.008sec$	2
4			$0.010sec$	4
5			$0.015sec$	6
6			$0.023sec$	9
7	$-0.62348980 + 0.78183148i$	$-2.68837595 + 3.37111682i$	$0.031sec$	12
8	$-0.70710678 - 0.70710678i$	$-3.69551813 - 3.69551813i$	$0.041sec$	16
9			$0.051sec$	20
10			$0.063sec$	25
11			$0.076sec$	30
12	$+1.00000000i$	$+9.46410162i$	$0.091sec$	36
13	$-0.97094182 - 0.23931566i$	$-10.34376769 - 2.54950976i$	$0.106sec$	42
14			$0.124sec$	49
15			$0.141sec$	56
16			$0.161sec$	64
17	$+0.27366299 + 0.96182564i$	$+4.34209501 + 15.26088102i$	$0.182sec$	72
18	-1.00000000	-17.27631145	$0.204sec$	81
19			$0.227sec$	90
20			$0.252sec$	100
21			$0.277sec$	110
22	$+0.41541501 + 0.90963200i$	$+9.68118116 + 21.19882974i$	$0.305sec$	121
23	$-0.99068595 + 0.13616665i$	$-24.67255764 + 3.39116499i$	$0.333sec$	132
24			$0.363sec$	144
25			$0.393sec$	156

TS-code:

```
fabcdekghijmlonqpsrut
mnsqhgfedrokaulcitpjb
pjilorsncbadhmtkfqueg
```

Table 24: Values of the quantum invariants for the 3-manifold induced by $5_1(1)$

			a single rigid 3-gem with 40 vertices $\pi_1 = \pi_1'$ homology sphere

r	$S^1 \times S^2 - normalization$	$S^3 - normalization$	$C^{++}, 486$ time	$r-$ states
3	+0.70710678	+1.00000000	0.007sec	2
4	−0.50000000	−1.00000000	0.009sec	4
5	−0.30075048 − 0.21850801i	−0.80901699 − 0.58778525i	0.014sec	6
6	+0.28867513	+1.00000000	0.021sec	9
7	−0.75871397 + 0.58875155i	−3.27143827 + 2.53859088i	0.028sec	12
8	+0.19134172i	+1.00000000i	0.037sec	16
9	−0.87249112	−5.41147413	0.046sec	20
10	+0.04270510 − 0.13143278i	+0.30901699 − 0.95105652i	0.058sec	25
11	+0.10106077 + 0.77058280i	+0.84125353 + 6.41451195i	0.069sec	30
12	−0.78867513 − 0.18301270i	−7.46410162 − 1.73205081i	0.083sec	36
13	−0.19359388 + 0.73407743i	−2.06242032 + 7.82037216i	0.096sec	42
14	−0.52375260 − 0.22974858i	−6.22736527 − 2.73168731i	0.112sec	49
15	−0.20503417 + 0.04818717i	−2.70071005 + 0.63472146i	0.128sec	56
16	−0.46450739 + 0.36696223i	−6.73444627 + 5.32023271i	0.147sec	64
17	+0.09191713 + 0.28501124i	+1.45841026 + 4.52215293i	0.165sec	72
18	−1.38180989	−23.87257811	0.185sec	81
19	+0.11031951 − 0.23139619i	+2.06584892 − 4.33313728i	0.206sec	90
20	−0.29120873 + 0.25600138i	−5.88670053 + 5.17499424i	0.229sec	100
21	−0.55006755 − 0.26011465i	−11.95917521 − 5.65522658i	0.252sec	110
22	−0.58411690 + 0.51013025i	−13.61275196 + 11.88850486i	0.277sec	121
23	−0.27290139 − 0.70279523i	−6.79647793 − 17.50277774i	0.302sec	132
24	−0.61190634 − 0.01379172i	−16.23969638 − 0.36602540i	0.329sec	144
25	−0.13031424 − 0.34060127i	−3.67604340 − 9.60804479i	0.357sec	156

TS-code:
$eabcdjfghilknmporqts$
$lqngfedmrkjpsciatohb$
$mihkrnqlbaocjtdgpsef$

Table 25: Values of the quantum invariants for the 3-manifold induced by $5_1(2)$

| | | | 235 rigid 3-gems with 40 vertices $\pi_1/\pi_1' = Z_2$ $\pi_1'/\pi_1'' = Z_5$ $\pi_1''/\pi_1''' = Z_4^4$ |

r	$S^1 \times S^2 - normalization$	$S^3 - normalization$	$C^{++}, 486$ time	$r-$ states
3			$0.007sec$	2
4	$+0.92387953i$	$+1.84775907i$	$0.008sec$	4
5			$0.013sec$	6
6	-0.78867513	-2.73205081	$0.019sec$	9
7			$0.025sec$	12
8	$+1.00367261 + 1.00367261i$	$+5.24544583 + 5.24544583i$	$0.033sec$	16
9			$0.042sec$	20
10	$-0.33659772 - 0.37348143i$	$-2.43564402 - 2.70253699i$	$0.052sec$	25
11			$0.062sec$	30
12	$+0.65328148 + 1.22569244i$	$+6.18272233 + 11.60007780i$	$0.074sec$	36
13			$0.087sec$	42
14	$+0.49681149 - 0.52940053i$	$+5.90703814 - 6.29451861i$	$0.101sec$	49
15			$0.116sec$	56
16	$-0.09404935 + 0.84296475i$	$-1.36353123 + 12.22133588i$	$0.132sec$	64
17			$0.149sec$	72
18	$+1.01214247 + 0.19746542i$	$+17.48608857 + 3.41147413i$	$0.167sec$	81
19			$0.186sec$	90
20	$-0.07915727 - 0.07709142i$	$-1.60014141 - 1.55838088i$	$0.206sec$	100
21			$0.227sec$	110
22	$+0.44005804 + 0.94767041i$	$+10.25548304 + 22.08530888i$	$0.249sec$	121
23			$0.272sec$	132
24	$+0.81711408 - 0.30383412i$	$+21.68581007 - 8.06360956i$	$0.297sec$	144
25			$0.322sec$	156

TS-code:
| $dabcgefjhimklpnorqts$ |
| $jqnsdmhgprtcflbioaek$ |
| $ptirlkbmqfjesdhacnog$ |

Table 26: Values of the quantum invariants for the 3-manifold induced by $5_1(3)$

			269 rigid 3-gems with 38 vertices $\pi_1/\pi_1' = Z_3$ $\pi_1'' = \pi_1'$	
r	$S^1 \times S^2$ – normalization	S^3 – normalization	$C^{++}, 486$ time	$r-$ states
3	$+0.70710678i$	$+1.00000000i$	$0.006sec$	2
4	$+0.35355339 - 0.35355339i$	$+0.70710678 - 0.70710678i$	$0.008sec$	4
5	$+0.22975292i$	$+0.61803399i$	$0.011sec$	6
6	$-0.50000000i$	$-1.73205081i$	$0.017sec$	9
7	$+1.17094801i$	$+5.04891734i$	$0.022sec$	12
8	$+0.17677670 - 0.42677670i$	$+0.92387953 - 2.23044250i$	$0.030sec$	16
9	$-0.35355339 - 0.20412415i$	$-2.19285330 - 1.26604444i$	$0.037sec$	20
10	$+0.05278640$	$+0.38196601$	$0.046sec$	25
11	$-0.48271312 - 0.16112635i$	$-4.01821720 - 1.34125353i$	$0.056sec$	30
12	$+0.70710678 + 0.35355339i$	$+6.69213043 + 3.34606521i$	$0.067sec$	36
13	$+0.34291117 - 0.22391638i$	$+3.65314731 - 2.38545603i$	$0.078sec$	42
14	-0.66029623	-7.85085508	$0.090sec$	49
15	$+0.30075048 + 0.09771975i$	$+3.96148528 + 1.28716460i$	$0.103sec$	56
16	$-0.73754462 + 0.15302135i$	$-10.69295076 + 2.21850946i$	$0.118sec$	64
17	$+0.65158532 + 0.07192364i$	$+10.33842895 + 1.14118207i$	$0.133sec$	72
18	$+0.25000000 - 0.72168784i$	$+4.31907786 - 12.46810383i$	$0.149sec$	81
19	$-0.39749685 + 0.48911577i$	$-7.44354700 + 9.15920785i$	$0.166sec$	90
20	$+0.17973259 - 0.26197506i$	$+3.63324264 - 5.29575039i$	$0.184sec$	100
21	$-0.32491356 + 0.67468996i$	$-7.06403823 + 14.66862652i$	$0.203sec$	110
22	$+0.28557744 - 0.06655686i$	$+6.65533713 - 1.55109708i$	$0.223sec$	121
23	$+0.03946506 - 0.80333998i$	$+0.98285833 - 20.00679646i$	$0.243sec$	132
24	$-0.25000000 + 0.35355339i$	$-6.63487833 + 9.38313491i$	$0.265sec$	144
25	$+0.12241491 - 0.13692955i$	$+3.45321080 - 3.86265525i$	$0.287sec$	156

TS-code:
dabcgef jhimklonqpsr
jlqedchgpoisnmarfkb
kpgiojrqlsadhbenmcf

Table 27: Values of the quantum invariants for the 3-manifold induced by $5_1(4)$

$$\begin{array}{c} \text{522 rigid 3-gems} \\ \text{with 36 vertices} \\ \pi_1/\pi_1' = Z_4 \\ \pi_1'/\pi_1'' = Z_5 \\ \pi_1''/\pi_1''' = Z_3^4 \end{array}$$

r	$S^1 \times S^2$ − normalization	S^3 − normalization	C^{++}, 486 time	$r-$ states
3	$+0.70710678 + 0.70710678i$	$+1.00000000 + 1.00000000i$	$0.006sec$	2
4	$-0.65328148 + 0.27059805i$	$-1.30656296 + 0.54119610i$	$0.007sec$	4
5	$-0.46842985 - 0.23867693i$	$-1.26007351 - 0.64203952i$	$0.012sec$	8
6	$+0.28867513 - 0.50000000i$	$+1.00000000 - 1.73205081i$	$0.019sec$	13
7	$-0.31443577 + 0.50042166i$	$-1.35579053 + 2.15772826i$	$0.028sec$	20
8	$+0.27778512 - 0.41573481i$	$+1.45177498 - 2.17273480i$	$0.041sec$	30
9	$-0.34923407 + 0.62849235i$	$-2.16606348 + 3.89811429i$	$0.057sec$	42
10	$-0.36180340 - 0.26286556i$	$-2.61803399 - 1.90211303i$	$0.076sec$	57
11	$+0.40300692 - 0.49472960i$	$+3.35472409 - 4.11824521i$	$0.100sec$	76
12	$-0.06393390 + 1.07822962i$	$-0.60507694 + 10.20447473i$	$0.127sec$	98
13	$-0.41578388 - 0.63193927i$	$-4.42948460 - 6.73226024i$	$0.159sec$	124
14	$+0.08410501 + 0.22252093i$	$+1.00000000 + 2.64575131i$	$0.197sec$	155
15	$+0.05455923 + 0.04302805i$	$+0.71865419 + 0.56676544i$	$0.239sec$	190
16	$+0.04251121 - 0.14014067i$	$+0.61632914 - 2.03176488i$	$0.288sec$	230
17	$-0.23508163 + 0.80100411i$	$-3.72994082 + 12.70919378i$	$0.343sec$	276
18	$-0.24465185 - 0.95896453i$	$-4.22668160 - 16.56736987i$	$0.405sec$	327
19	$+0.52616909 + 0.51533471i$	$+9.85307019 + 9.65018506i$	$0.473sec$	384
20	$-0.28608593 + 0.39541300i$	$-5.78314462 + 7.99316000i$	$0.549sec$	448
21	$-0.31277605 - 0.53567527i$	$-6.80015312 - 11.64626855i$	$0.633sec$	518
22	$+0.28929802 + 0.51193923i$	$+6.74204451 + 11.93066266i$	$0.725sec$	595
23	$-0.08255371 - 0.49029725i$	$-2.05596061 - 12.21061762i$	$0.827sec$	680
24	$+0.11949230 + 0.68072938i$	$+3.17126740 + 18.06622649i$	$0.935sec$	772
25	$-0.43208162 - 0.06699411i$	$-12.18862036 - 1.88984153i$	$1.055sec$	872

TS-code:
$$\begin{array}{c} dabcgefjhimklpnorq \\ jofedcqgmahnplrikb \\ ohmqrknbafjecgidpl \end{array}$$

Table 28: Values of the quantum invariants for the 3-manifold induced by $5_1(5)$

r	$S^1 \times S^2$ − normalization	S^3 − normalization	C^{++}, 486 time	$r-$ states
3	+0.70710678	+1.00000000	0.005sec	2
4	+0.50000000i	+1.00000000i	0.006sec	4
5			0.009sec	6
6	−0.28867513	−1.00000000	0.013sec	9
7	−0.67846695 − 0.77894422i	−2.92542754 − 3.35866748i	0.017sec	12
8	+0.59723879 − 0.59723879i	+3.12132034 − 3.12132034i	0.023sec	16
9	+0.39335126 + 0.40204608i	+2.43969262 + 2.49362077i	0.028sec	20
10			0.035sec	25
11	−0.43751771 + 0.03384486i	−3.64200004 + 0.28173256i	0.044sec	30
12	−0.50000000 − 0.89433757i	−4.73205081 − 8.46410162i	0.052sec	36
13	+0.87667262 − 0.38922509i	+9.33948646 − 4.14654496i	0.059sec	42
14	+0.30681073 + 0.44325679i	+3.64794847 + 5.27027818i	0.069sec	49
15			0.079sec	56
16	−0.49658591 − 0.11249703i	−7.19952191 − 1.63098631i	0.091sec	64
17	+0.01807421 − 0.97261592i	+0.28677583 − 15.43208579i	0.102sec	72
18	+1.00948397 + 0.12088216i	+17.44015945 + 2.08839786i	0.115sec	81
19	+0.08204441 + 0.44920298i	+1.53636792 + 8.41179870i	0.127sec	90
20			0.141sec	100
21	−0.40432911 − 0.30932987i	−8.79063424 − 6.72522889i	0.155sec	110
22	+0.57660557 − 0.70012510i	+13.43770156 − 16.31630476i	0.171sec	121
23	+0.75717162 + 0.67599010i	+18.85699558 + 16.83520880i	0.187sec	132
24	−0.14909074 + 0.31284903i	−3.95679568 + 8.30286089i	0.204sec	144
25			0.220sec	156

TS-code:
```
dabcgefjhilknmpo
jmoedchgpniblafk
ejimpdkocbgnfhla
```

Table 29: Values of the quantum invariants for the 3-manifold induced by $5_1(6)$

$$R_{30}(335) \in C_{30}(28)$$
TS-class with 23 rigid 3-gems
$$\pi_1/\pi_1' = Z_6$$
$$\pi_1'/\pi_1'' = Z_5$$
$$\pi_1''/\pi_1''' = Z_2^4$$

r	$S^1 \times S^2$ – normalization	S^3 – normalization	C^{++}, 486 time	$r-$ states
3			$0.006 sec$	2
4	$+0.27059805 + 0.27059805i$	$+0.54119610 + 0.54119610i$	$0.007 sec$	4
5			$0.012 sec$	8
6	$+0.50000000 - 0.50000000i$	$+1.73205081 - 1.73205081i$	$0.019 sec$	13
7			$0.028 sec$	20
8	$+0.08515863 + 0.71538670i$	$+0.44506044 + 3.73879105i$	$0.041 sec$	30
9			$0.057 sec$	42
10	$+0.33659772 - 0.03594128i$	$+2.43564402 - 0.26007351i$	$0.076 sec$	57
11			$0.100 sec$	76
12	$-0.73253782 + 0.46193977i$	$-6.93281233 + 4.37184489i$	$0.126 sec$	98
13			$0.159 sec$	124
14	$+0.21397183 - 0.16399264i$	$+2.54410338 - 1.94985582i$	$0.197 sec$	155
15			$0.239 sec$	190
16	$-0.60418383 - 0.44892639i$	$-8.75948084 - 6.50855236i$	$0.288 sec$	230
17			$0.343 sec$	276
18	$+0.60566243 + 0.10566243i$	$+10.46361282 + 1.82545710i$	$0.404 sec$	327
19			$0.473 sec$	384
20	$+0.06390936 - 0.44217489i$	$+1.29190928 - 8.93843813i$	$0.549 sec$	448
21			$0.633 sec$	518
22	$+0.12984061 + 0.78953486i$	$+3.02591476 + 18.39998509i$	$0.725 sec$	595
23			$0.827 sec$	680
24	$+0.06897484 - 0.24117784i$	$+1.83055881 - 6.40074249i$	$0.934 sec$	772
25			$1.055 sec$	872

TS-code:
$$dabcgefjhimklon$$
$$jloedchgnmibakf$$
$$fohmladcbkjngei$$

Table 30: Values of the quantum invariants for the 3-manifold induced by $5_1(7)$

$$\equiv \quad \xrightarrow{TS_\rho} \quad 7$$

$R_{26}(10) \in C_{26}(13)$
TS-class with a single 3-gem
$\pi_1/\pi_1' = Z_7$
$\pi_1'' = \pi_1'$
3-manifold: $S^3/(\langle \bar{5},3,2\rangle \times Z_7)$

r	$S^1 \times S^2$ − normalization	S^3 − normalization	$C^{++}, 486$ time	$r-$ states
3	+0.70710678i	+1.00000000i	0.006sec	2
4	−0.35355339 − 0.35355339i	−0.70710678 − 0.70710678i	0.008sec	4
5	−0.21850801 − 0.07099756i	−0.58778525 − 0.19098301i	0.011sec	6
6	+0.28867513	+1.00000000	0.017sec	9
7	−0.68937814 + 0.15734606i	−2.97247462 + 0.67844793i	0.022sec	12
8	+0.67677670 − 0.28033009i	+3.53700546 − 1.46507563i	0.030sec	16
9	−0.26241694 + 0.55975478i	−1.62759536 + 3.47178151i	0.037sec	20
10	−0.04270510 + 0.03102707i	−0.30901699 + 0.22451399i	0.046sec	25
11	+0.06494781 − 0.48892915i	+0.54064082 − 4.06996095i	0.056sec	30
12	+0.33353367 + 0.63239216i	+3.15659652 + 5.98502365i	0.067sec	36
13	−0.30327274 − 0.62336697i	−3.23086585 − 6.64093665i	0.078sec	42
14	+0.39091574 + 0.31174490i	+4.64794847 + 3.70661522i	0.090sec	49
15	−0.14320045 − 0.06445060i	−1.88623634 − 0.84894331i	0.103sec	56
16	−0.44606695 − 0.21509707i	−6.46709609 − 3.11848568i	0.118sec	64
17	+0.73481939 − 0.14051572i	+11.65906874 − 2.22950356i	0.133sec	72
18	−0.71449816 + 0.37152172i	−12.34389279 + 6.41852487i	0.149sec	81
19	+0.27090101 − 0.28224845i	+5.07290656 − 5.28539933i	0.166sec	90
20	−0.08618550 + 0.08998111i	−1.74221509 + 1.81894222i	0.184sec	100
21	−0.17713229 + 0.36781900i	−3.85108366 + 7.99685762i	0.203sec	110
22	+0.14342254 − 0.67722979i	+3.34245703 − 15.78273311i	0.223sec	121
23	+0.21375515 + 0.85898038i	+5.32346941 + 21.39249379i	0.243sec	132
24	−0.28481147 − 0.39892058i	−7.55875786 − 10.58715816i	0.265sec	144
25	+0.10429884 − 0.03567316i	+2.94217335 − 1.00630676i	0.288sec	156

TS-code:	$cabfdeighkjml$
	$imdcbkfjalghe$
	$ldfbjhmceiakg$

Table 31: Values of the quantum invariants for the 3-manifold induced by $5_1(8)$

$$R_{24}(4) \in C_{24}(14)$$
TS-class with a single 3-gem
$$\pi_1/\pi_1' = Z_8$$
3-manifold: $S^3/(Z_5 \rtimes_i Z_8)$

r	$S^1 \times S^2 - normalization$	$S^3 - normalization$	$C^{++}, 486$ time	$r-$ states
3	$+0.70710678 + 0.70710678i$	$+1.00000000 + 1.00000000i$	$0.007sec$	2
4	$-0.38268343 + 0.92387953i$	$-0.76536686 + 1.84775907i$	$0.008sec$	4
5	$-0.22975292 + 0.22975292i$	$-0.61803399 + 0.61803399i$	$0.013sec$	6
6	$-0.28867513 + 0.50000000i$	$-1.00000000 + 1.73205081i$	$0.019sec$	9
7	$-1.34746553 - 0.16996242i$	$-5.81002915 - 0.73284740i$	$0.025sec$	12
8	$-0.27589938 - 1.38703985i$	$-1.44191964 - 7.24901957i$	$0.033sec$	16
9	$+1.03939610 - 0.51456222i$	$+6.44667323 - 3.19148250i$	$0.042sec$	20
10	$+0.55278640$	$+4.00000000$	$0.052sec$	25
11	$+0.82933635 + 0.17212458i$	$+6.90359028 + 1.43280537i$	$0.062sec$	30
12	$+0.30732410 + 1.49629049i$	$+2.90854651 + 14.16104524i$	$0.075sec$	36
13	$-1.30928196 + 0.75415770i$	$-13.94821831 + 8.03429404i$	$0.087sec$	42
14	$-0.81184821 - 0.75500169i$	$-9.65279281 - 8.97689340i$	$0.101sec$	49
15	$-0.28686458 - 0.51661750i$	$-3.77858026 - 6.80488572i$	$0.116sec$	56
16	$-0.09801714 - 0.99518473i$	$-1.42105634 - 14.42822712i$	$0.132sec$	64
17	$+1.43385083 - 0.66988891i$	$+22.75030515 - 10.62884417i$	$0.149sec$	72
18	$+0.99757535 + 1.09881447i$	$+17.23442238 + 18.98346096i$	$0.167sec$	81
19	$-0.49989842 + 0.85457535i$	$-9.36112418 + 16.00282312i$	$0.186sec$	90
20	$-0.35900605 + 0.42034208i$	$-7.25720394 + 8.49709409i$	$0.206sec$	100
21	$-0.96299110 + 0.32391336i$	$-20.93666366 + 7.04229257i$	$0.227sec$	110
22	$-0.83897085 - 1.24533599i$	$-19.55208302 - 29.02235823i$	$0.250sec$	121
23	$+0.88512569 - 1.01185801i$	$+22.04363028 - 25.19983775i$	$0.272sec$	132
24	$+0.73229353 + 0.36112720i$	$+19.43471380 + 9.58414022i$	$0.297sec$	144
25	$+0.40030708 + 0.17570712i$	$+11.29229040 + 4.95653442i$	$0.322sec$	156

TS-code:

```
cabfdeighljk
ildcbkfjahge
kfhaibjlegdc
```

Table 32: Values of the quantum invariants for the 3-manifold induced by $5_1(9)$

$$\equiv \qquad 9 \quad \xrightarrow{TS_\rho}$$

$$\boxed{\begin{array}{c} R_{24}(14) \in C_{24}(17) \\ TS\text{-class with 14 rigid 3-gems.} \\ \pi_1/\pi_1' = Z_9 \\ 3\text{-manifold: } L_{9,2} \end{array}}$$

r	$S^1 \times S^2 - normalization$	$S^3 - normalization$	C^{++}, 486 time	$r-$ states
3	$+0.70710678$	$+1.00000000$	$0.007sec$	2
4	$+0.50000000$	$+1.00000000$	$0.009sec$	4
5	$-0.11487646 - 0.35355339i$	$-0.30901699 - 0.95105652i$	$0.014sec$	6
6	$-0.50000000i$	$-1.73205081i$	$0.021sec$	9
7	$+0.11596031 - 0.50805530i$	$+0.50000000 - 2.19064313i$	$0.028sec$	12
8	$+0.19134172$	$+1.00000000$	$0.037sec$	16
9			$0.046sec$	20
10	$-0.11180340 - 0.08122992i$	$-0.80901699 - 0.58778525i$	$0.058sec$	25
11	$+0.17533059 - 0.38392045i$	$+1.45949297 - 3.19584383i$	$0.069sec$	30
12	$+0.25000000 - 0.25000000i$	$+2.36602540 - 2.36602540i$	$0.083sec$	36
13	$+0.23030543 - 0.12087362i$	$+2.45352077 - 1.28770704i$	$0.096sec$	42
14	$-0.14692973 - 0.18424406i$	$-1.74697960 - 2.19064313i$	$0.112sec$	49
15	$-0.09771975 - 0.30075048i$	$-1.28716460 - 3.96148528i$	$0.128sec$	56
16	$-0.34675996i$	$-5.02733949i$	$0.147sec$	64
17	$+0.12179515 - 0.02276746i$	$+1.93247223 - 0.36124167i$	$0.165sec$	72
18			$0.185sec$	81
19	$-0.09264941 - 0.05013935i$	$-1.73495775 - 0.93891218i$	$0.206sec$	90
20	$+0.09651666 - 0.29704774i$	$+1.95105652 - 6.00473452i$	$0.229sec$	100
21	$+0.16663466 - 0.20895325i$	$+3.62285157 - 4.54291217i$	$0.252sec$	110
22	$+0.16610378 - 0.10674842i$	$+3.87102225 - 2.48775494i$	$0.277sec$	121
23	$-0.12702156 - 0.13600688i$	$-3.16341089 - 3.38718599i$	$0.302sec$	132
24	$-0.09567086 - 0.23096988i$	$-2.53905801 - 6.12982829i$	$0.329sec$	144
25	$-0.01744527 - 0.27728455i$	$-0.49211470 - 7.82193919i$	$0.357sec$	156

$$TS\text{-code:} \quad \boxed{\begin{array}{l} cabfdeighljk \\ ilkcjgfeahdb \\ lkigbhdfjcea \end{array}}$$

Table 33: Values of the quantum invariants for the 3-manifold induced by $5_1(10)$

$$R_{22}(5) \in C_{22}(9)$$
TS-class with 3 rigid
3-gems.
$$\pi_1/\pi_1' = Z_{10}$$
3-manifold: $L_{5,2}\#L_{2,1}$

r	$S^1 \times S^2$ − *normalization*	S^3 − *normalization*	C^{++}, 486 time	$r-$ states
3			0.008sec	2
4	−0.38268343	−0.76536686	0.010sec	4
5			0.015sec	6
6	−0.21132487	−0.73205081	0.023sec	9
7			0.031sec	12
8	+0.33304001	+1.74055098	0.041sec	16
9			0.051sec	20
10			0.063sec	25
11			0.076sec	30
12	−0.28124486	−2.66172990	0.091sec	36
13			0.106sec	42
14	+0.16768905	+1.99380581	0.124sec	49
15			0.141sec	56
16	+0.13956460	+2.02341303	0.161sec	64
17			0.182sec	72
18	−0.22233372	−3.84110662	0.204sec	81
19			0.227sec	90
20			0.251sec	100
21			0.277sec	110
22	+0.20508713	+4.77952307	0.305sec	121
23			0.333sec	132
24	−0.12452953	−3.30495321	0.362sec	144
25			0.392sec	156

TS-code:
dabcgefihkj
ihkedjbgafc
higkbdjaecf

Table 34: Values of the quantum invariants for the 3-manifold induced by $5_1(11)$

$$R_{24}(22) \in C_{24}(20)$$
TS-class with 10 rigid 3-gems.
$$\pi_1/\pi_1' = Z_{11}$$
3-manifold: $L_{11,3}$

r	$S^1 \times S^2$ − normalization	S^3 − normalization	$C^{++}, 486$ time	$r-$ states
3	$+0.70710678i$	$+1.00000000i$	$0.008sec$	2
4	$-0.35355339 + 0.35355339i$	$-0.70710678 + 0.70710678i$	$0.011sec$	4
5	$-0.21850801 + 0.30075048i$	$-0.58778525 + 0.80901699i$	$0.017sec$	6
6	-0.28867513	-1.00000000	$0.025sec$	9
7	$-0.40742872 + 0.09299295i$	$-1.75675939 + 0.40096887i$	$0.033sec$	12
8	$-0.42677670 - 0.17677670i$	$-2.23044250 - 0.92387953i$	$0.044sec$	16
9	$-0.46424283i$	$-2.87938524i$	$0.055sec$	20
10	$+0.04270510 - 0.13143278i$	$+0.30901699 - 0.95105652i$	$0.069sec$	25
11			$0.083sec$	30
12	$-0.07471462 - 0.07471462i$	$-0.70710678 - 0.70710678i$	$0.099sec$	36
13	$-0.18095040 - 0.34477219i$	$-1.92772512 - 3.67297334i$	$0.116sec$	42
14	$+0.21231962 - 0.26624038i$	$+2.52445867 - 3.16557105i$	$0.134sec$	49
15	$+0.25807691 - 0.08385427i$	$+3.39938907 - 1.10452846i$	$0.154sec$	56
16	$+0.16332037 - 0.10912718i$	$+2.36782513 - 1.58213017i$	$0.176sec$	64
17	$+0.17367180 + 0.04941389i$	$+2.75557705 + 0.78402927i$	$0.197sec$	72
18	$+0.25534815$	$+4.41147413$	$0.222sec$	81
19	$+0.24873564 + 0.16250710i$	$+4.65783675 + 3.04311663i$	$0.247sec$	90
20	$-0.04885988 + 0.30848912i$	$-0.98768834 + 6.23601876i$	$0.274sec$	100
21	$-0.03946755 + 0.08195521i$	$-0.85807523 + 1.78181158i$	$0.301sec$	110
22			$0.331sec$	121
23	$+0.05020855 + 0.06171460i$	$+1.25041991 + 1.53697255i$	$0.361sec$	132
24	$+0.10952609 + 0.26441938i$	$+2.90676926 + 7.01756176i$	$0.394sec$	144
25	$-0.17519185 + 0.18656038i$	$-4.94199920 + 5.26269462i$	$0.427sec$	156

TS-code:

dabcgefjhilk
jikedclgbafh
ehlfidkbagjc

Table 35: Values of the quantum invariants for the 3-manifold induced by $5_1(12)$

$$R_{26}(6) \in C_{26}(24)$$
$T S$-class with 4 rigid 3-gems.
$$\pi_1/\pi_1' = Z_{12}$$
3-manifold: $S^3/(Z_5 \rtimes_i Z_{12})$

r	$S^1 \times S^2$ − normalization	S^3 − normalization	C^{++}, 486 time	$r-$ states
3	$+0.70710678 + 0.70710678i$	$+1.00000000 + 1.00000000i$	$0.009sec$	2
4	$+0.27059805 + 0.65328148i$	$+0.54119610 + 1.30656296i$	$0.012sec$	4
5	$-0.14751046 - 0.28950557i$	$-0.39680225 - 0.77876826i$	$0.018sec$	6
6	$-1.00000000i$	$-3.46410162i$	$0.027sec$	9
7	$+0.10047727 - 1.45741117i$	$+0.43323994 - 6.28409501i$	$0.036sec$	12
8	$+0.13983543 - 1.10925473i$	$+0.73081516 - 5.79724459i$	$0.048sec$	16
9	$-0.14942925 - 0.55767754i$	$-0.92680886 - 3.45889774i$	$0.060sec$	20
10	$-0.44721360 + 0.32491970i$	$-3.23606798 + 2.35114101i$	$0.075sec$	25
11	$-0.66952203 + 0.87164357i$	$-5.57325842 + 7.25576549i$	$0.090sec$	30
12	$-0.54119610 + 1.30656296i$	$-5.12193489 + 12.36544467i$	$0.107sec$	36
13	$-0.21207665 + 1.19662730i$	$-2.25932341 + 12.74807052i$	$0.125sec$	42
14	$+0.23565699 + 0.88651357i$	$+2.80193774 + 10.54055637i$	$0.146sec$	49
15	$+0.54598069 + 0.08647485i$	$+7.19165759 + 1.13904666i$	$0.167sec$	56
16	$+0.72624364 - 0.67165733i$	$+10.52910870 - 9.73771425i$	$0.190sec$	64
17	$+0.69474126 - 1.38634844i$	$+11.02316602 - 21.99660476i$	$0.214sec$	72
18	$+0.50000000 - 1.44337567i$	$+8.63815572 - 24.93620766i$	$0.240sec$	81
19	$+0.08054409 - 1.06656254i$	$+1.50827288 - 19.97250661i$	$0.267sec$	90
20	$-0.42856110 - 0.14801415i$	$-8.66323932 - 2.99206340i$	$0.297sec$	100
21	$-0.89127685 + 0.65935624i$	$-19.37750378 + 14.33525167i$	$0.327sec$	110
22	$-1.04298146 + 1.29452487i$	$-24.30651798 + 30.16869738i$	$0.359sec$	121
23	$-0.80246532 + 1.29945912i$	$-19.98501336 + 32.36240531i$	$0.392sec$	132
24	$-0.19509032 + 0.98078528i$	$-5.17760220 + 26.02956400i$	$0.427sec$	144
25	$+0.49367554 + 0.25495612i$	$+13.92612771 + 7.19207490i$	$0.463sec$	156

$T S$-code:

```
cabfdeighkjml
imdcbgflkhaje
gfmaibdklchej
```

Table 36: Values of the quantum invariants for the 3-manifold induced by $5_1(13)$

$$R_{30}(253) \in C_{30}(56)$$
TS-class with 38 rigid 3-gems
$$\pi_1/\pi_1' = Z_{13}$$
$$\pi_1' = \pi_1''$$

r	$S^1 \times S^2$ − normalization	S^3 − normalization	$C^{++}, 486$ time	$r-$ states
3	+0.70710678	+1.00000000	0.009sec	2
4	−0.50000000i	−1.00000000i	0.013sec	4
5	−0.22975292	−0.61803399	0.020sec	6
6	+0.28867513	+1.00000000	0.029sec	9
7	+0.23192061 − 0.81485743i	+1.00000000 − 3.51351879i	0.039sec	12
8	−0.59723879 − 0.59723879i	−3.12132034 − 3.12132034i	0.051sec	16
9	−0.48886321 + 0.13962914i	−3.03208889 + 0.86602540i	0.064sec	20
10	+0.05278640	+0.38196601	0.080sec	25
11	−0.27639136 + 0.13739448i	−2.30074651 + 1.14370386i	0.096sec	30
12	−0.18301270 + 0.78867513i	−1.73205081 + 7.46410162i	0.115sec	36
13	+0.58193747 + 0.40168243i	+6.19957439 + 4.27925723i	0.135sec	42
14	+0.37796447 − 0.16399264i	+4.49395921 − 1.94985582i	0.157sec	49
15	+0.00896098 + 0.22473227i	+0.11803399 + 2.96017343i	0.179sec	56
16	+0.45925699 − 0.02237705i	+6.65832581 − 0.32442335i	0.205sec	64
17	+0.12833413 − 0.85708347i	+2.03622339 − 13.59898127i	0.230sec	72
18	−0.72640678 − 0.39214805i	−12.54962985 − 6.77487192i	0.259sec	81
19	−0.34212211 + 0.28003418i	−6.40659669 + 5.24393483i	0.288sec	90
20	+0.03594128 − 0.11930115i	+0.72654253 − 2.41163848i	0.319sec	100
21	−0.45018619 + 0.02213610i	−9.78762618 + 0.48126726i	0.351sec	110
22	−0.04547777 + 0.64854515i	−1.05985224 + 15.11424216i	0.386sec	121
23	+0.71807703 + 0.17916123i	+17.88336361 + 4.46192442i	0.421sec	132
24	+0.26670105 − 0.23291838i	+7.07811602 − 6.18154055i	0.460sec	144
25	+0.03489308 + 0.13956331i	+0.98430118 + 3.93695110i	0.498sec	156

TS-code:
dabcgefjhimklon
jlkedcngmaobihf
ohelcbmfanjdgki

14.5 Knot 5_2

Table 37: Values of the quantum invariants for the 3-manifold induced by $5_2(-6)$

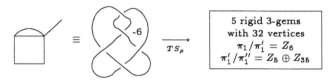

$$\begin{array}{c} \text{5 rigid 3-gems} \\ \text{with 32 vertices} \\ \pi_1/\pi_1' = Z_6 \\ \pi_1'/\pi_1'' = Z_5 \oplus Z_{35} \end{array}$$

r	$S^1 \times S^2 - normalization$	$S^3 - normalization$	$C^{++}, 486$ time	$r-$ states
3			$0.006sec$	2
4	$-0.65328148 - 0.65328148i$	$-1.30656296 - 1.30656296i$	$0.010sec$	6
5			$0.018sec$	10
6	$+0.50000000 + 0.50000000i$	$+1.73205081 + 1.73205081i$	$0.035sec$	19
7			$0.052sec$	28
8	$-0.46852587 - 0.70590645i$	$-2.44863418 - 3.68924492i$	$0.081sec$	44
9			$0.113sec$	60
10	$+0.07373217 + 0.77307572i$	$+0.53353098 + 5.59402846i$	$0.160sec$	85
11			$0.209sec$	110
12	$+0.05126986 - 0.60201162i$	$+0.48522315 - 5.69749918i$	$0.278sec$	146
13			$0.350sec$	182
14	$-0.11200731 + 0.70970924i$	$-1.33175551 + 8.43837080i$	$0.445sec$	231
15			$0.544sec$	280
16	$+0.41016954 - 0.68414742i$	$+5.94665411 - 9.91879610i$	$0.669sec$	344
17			$0.798sec$	408
18	$-0.47506776 + 0.47930054i$	$-8.20741860 + 8.28054534i$	$0.956sec$	489
19			$1.122sec$	570
20	$+0.49387652 - 0.49814635i$	$+9.98357168 - 10.06988504i$	$1.319sec$	670
21			$1.523sec$	770
22	$-0.70193612 + 0.40038967i$	$-16.35851021 + 9.33101782i$	$1.763sec$	891
23			$2.012sec$	1012

TS-code:
$$\begin{array}{l} dabcgefjhilknmpo \\ gpindjamkochlefb \\ kinplchgmajobfed \end{array}$$

Table 38: Values of the quantum invariants for the 3-manifold induced by $5_2(-5)$

r	$S^1 \times S^2$ − *normalization*	S^3 − *normalization*	$C^{++}, 486$ *time*	$r-$ *states*
3	+0.70710678	+1.00000000	0.006*sec*	2
4	+0.50000000i	+1.00000000i	0.009*sec*	6
5	+0.25687157 + 0.35355339i	+0.69098301 + 0.95105652i	0.016*sec*	10
6	−0.28867513	−1.00000000	0.031*sec*	19
7	−0.23759308 + 0.58875155i	−1.02445867 + 2.53859088i	0.047*sec*	28
8	+0.05604269 − 0.05604269i	+0.29289322 − 0.29289322i	0.073*sec*	44
9	−0.57802870 − 0.23076559i	−3.58512231 − 1.43128334i	0.102*sec*	60
10	−0.05901699 − 0.18163563i	−0.42705098 − 1.31432778i	0.145*sec*	85
11	−0.17890674 − 0.72575072i	−1.48926169 − 6.04131920i	0.192*sec*	110
12	−0.18301270 − 0.07735027i	−1.73205081 − 0.73205081i	0.254*sec*	146
13	+0.44783276 − 0.39914726i	+4.77091205 − 4.25224921i	0.319*sec*	182
14	−0.05243861 − 0.52525311i	−0.62348980 − 6.24520610i	0.406*sec*	231
15	+0.41927136 − 0.08348209i	+5.52264232 − 1.09962613i	0.497*sec*	280
16	+0.40954154 − 0.46305424i	+5.93754928 − 6.71337851i	0.611*sec*	344
17	+0.13121136 + 0.28673378i	+2.08187525 + 4.54948373i	0.729*sec*	408
18	+0.66708770 + 0.23040125i	+11.52481493 + 3.98048369i	0.875*sec*	489
19	+0.01878787 + 0.03117295i	+0.35182260 + 0.58374623i	1.027*sec*	570
20	+0.18869068 + 0.54930655i	+3.81432778 + 11.10407376i	1.208*sec*	670
21	+0.13578693 + 0.06783508i	+2.95218236 + 1.47482169i	1.394*sec*	770
22	−0.39396192 + 0.56726340i	−9.18122012 + 13.21998388i	1.616*sec*	891
23	+0.09122886 + 0.45150261i	+2.27201085 + 11.24445571i	1.843*sec*	1012
24	−0.50237503 − 0.06117831i	−13.33278876 − 1.62364244i	2.106*sec*	1156
25	−0.38594801 + 0.31951437i	−10.88723421 + 9.01320317i	2.377*sec*	1300

TS-code:

    cabfdeighljknmpo
    ipjclgfendohkamb
    hengmolpjicdbkfa

Table 39: Values of the quantum invariants for the 3-manifold induced by $5_2(-4)$

$$R_{30}(170) \in C_{30}(18)$$
TS-class with 11 rigid 3-gems
$$\pi_1/\pi_1' = Z_4$$
$$\pi_1'/\pi_1'' = Z_3 \oplus Z_{21}$$

r	$S^1 \times S^2$ – normalization	S^3 – normalization	C^{++}, 486 time	$r-$ states
3	$+0.70710678 - 0.70710678i$	$+1.00000000 - 1.00000000i$	$0.006sec$	2
4	$+0.27059805 - 0.65328148i$	$+0.54119610 - 1.30656296i$	$0.010sec$	6
5	$-0.74901141 - 0.31199539i$	$-2.01483624 - 0.83926573i$	$0.018sec$	10
6	$+0.28867513 + 0.50000000i$	$+1.00000000 + 1.73205081i$	$0.034sec$	19
7	$+0.04377363 - 0.31685322i$	$+0.18874401 - 1.36621412i$	$0.052sec$	28
8	$-0.43325201 - 0.38481052i$	$-2.26428411 - 2.01111669i$	$0.081sec$	44
9	$-0.76202269 + 1.17702475i$	$-4.72631299 + 7.30029094i$	$0.113sec$	60
10	$+1.13819660 + 0.62613682i$	$+8.23606798 + 4.53076859i$	$0.160sec$	85
11	$+0.84875173 - 0.51724069i$	$+7.06520848 - 4.30563282i$	$0.209sec$	110
12	$-0.69554619 - 0.85994563i$	$-6.58271984 - 8.13861284i$	$0.278sec$	146
13	$-0.08800762 + 0.78366086i$	$-0.93757459 + 8.34860105i$	$0.351sec$	182
14	$+0.36871996 - 0.22252093i$	$+4.38404294 - 2.64575131i$	$0.445sec$	231
15	$-0.26611347 - 0.73128916i$	$-3.50524659 - 9.63254076i$	$0.544sec$	280
16	$-1.24118112 + 0.92284887i$	$-17.99469252 + 13.37949899i$	$0.669sec$	344
17	$+0.98343263 + 1.02460052i$	$+15.60371001 + 16.25690352i$	$0.798sec$	408
18	$+1.08997161 - 0.39848223i$	$+18.83068904 - 6.88430305i$	$0.957sec$	489
19	$-0.60849415 - 1.11264987i$	$-11.39469364 - 20.83554034i$	$1.121sec$	570
20	$-0.31862371 + 0.82240579i$	$-6.44088647 + 16.62469620i$	$1.319sec$	670
21	$+0.47127313 - 0.08724784i$	$+10.24608334 - 1.89688023i$	$1.523sec$	770
22	$-0.11254585 - 0.85500805i$	$-2.62286312 - 19.92582742i$	$1.762sec$	891
23	$-1.42138075 + 0.73593460i$	$-35.39880494 + 18.32809762i$	$2.011sec$	1012
24	$+0.85356006 + 1.19220898i$	$+22.65306865 + 31.64064605i$	$2.297sec$	1156
25	$+1.20441994 - 0.30328960i$	$+33.97556595 - 8.55551754i$	$2.593sec$	1300

TS-code:
```
dabcgefjhimklon
jkmedchglobnfia
nflkhbomgdjceai
```

Table 40: Values of the quantum invariants for the 3-manifold induced by $5_2(-3)$

$$R_{28}(2) \in C_{28}(7)$$
$$TS\text{-class with a single rigid 3-gem}$$
$$\pi_1/\pi_1' = Z_3$$
$$\pi_1'/\pi_1'' = Z_5 \oplus Z_5$$
$$\pi_1''/\pi_1''' = S_6^5$$

r	$S^1 \times S^2$ − normalization	S^3 − normalization	$C^{++}, 486$ time	$r-$ states
3	$-0.70710678i$	$-1.00000000i$	$0.007 sec$	2
4	$-0.35355339 - 0.35355339i$	$-0.70710678 - 0.70710678i$	$0.012 sec$	6
5	$+0.13504538 + 0.18587402i$	$+0.36327126 + 0.50000000i$	$0.020 sec$	10
6	$+0.50000000i$	$+1.73205081i$	$0.038 sec$	19
7	$-0.10062658 - 0.73007414i$	$-0.43388374 - 3.14794847i$	$0.058 sec$	28
8	$-0.42677670 + 0.17677670i$	$-2.23044250 + 0.92387953i$	$0.090 sec$	44
9	$+0.66446302 + 0.38362791i$	$+4.12121613 + 2.37938524i$	$0.125 sec$	60
10	$+0.01631190 + 0.05020285i$	$+0.11803399 + 0.36327126i$	$0.177 sec$	85
11	$-0.45323982 - 0.53732629i$	$-3.77287459 - 4.47283007i$	$0.232 sec$	110
12	$+0.35355339i$	$+3.34606521i$	$0.308 sec$	146
13	$+0.66385087 - 0.28872185i$	$+7.07222514 - 3.07585041i$	$0.388 sec$	182
14	$-0.55747609 + 0.24598896i$	$-6.62833414 + 2.92478374i$	$0.492 sec$	231
15	$-0.27673476 + 0.13077456i$	$-3.64515022 + 1.72256245i$	$0.600 sec$	280
16	$+0.48996111 - 0.02617184i$	$+7.10347538 - 0.37944039i$	$0.738 sec$	344
17	$+0.15243122 - 0.63539756i$	$+2.41856172 - 10.08158459i$	$0.881 sec$	408
18	$-0.35286853 + 0.78107901i$	$-6.09626666 + 13.49416426i$	$1.057 sec$	489
19	$+0.08353579 - 0.13115072i$	$+1.56429571 - 2.45593527i$	$1.237 sec$	570
20	$-0.06138127 - 0.38754606i$	$-1.24080462 - 7.83413206i$	$1.455 sec$	670
21	$-0.14719050 - 0.05137243i$	$-3.20011056 - 1.11690264i$	$1.679 sec$	770
22	$+0.51457322 + 0.78507066i$	$+11.99204758 + 18.29594764i$	$1.944 sec$	891
23	$-0.18670214 - 0.60057694i$	$-4.64972720 - 14.95708028i$	$2.215 sec$	1012
24	-0.43301270	-11.49194636	$2.530 sec$	1156

TS-code:
$$\begin{array}{l} cabfdeighljknm \\ imdclgfkabnejh \\ gnejhblcdimakf \end{array}$$

Table 41: Values of the quantum invariants for the 3-manifold induced by $5_2(-2)$

$$R_{30}(1179) \in C_{30}(8)$$
$$TS\text{-class with 2 rigid 3-gems}$$
$$\pi_1/\pi_1' = Z_2$$
$$\pi_1'/\pi_1'' = Z_7$$
$$\pi_1''/\pi_1''' = Z_2^6$$

r	$S^1 \times S^2$ – normalization	S^3 – normalization	$C^{++}, 486$ time	$r-$ states
3			0.007sec	2
4	-0.38268343	-0.76536686	0.013sec	6
5			0.023sec	10
6	-0.78867513	-2.73205081	0.043sec	19
7			0.064sec	28
8	$-0.69351992i$	$-3.62450979i$	0.100sec	44
9			0.138sec	60
10	$-0.24250224 - 0.44184580i$	$-1.75476272 - 3.19722621i$	0.196sec	85
11			0.256sec	110
12	$+0.57792215 - 0.65328148i$	$+5.46951396 - 6.18272233i$	0.340sec	146
13			0.427sec	182
14	$+0.22075995 - 0.17830257i$	$+2.62481339 - 2.11999943i$	0.542sec	231
15			0.661sec	280
16	$+0.83887233 - 0.27746929i$	$+12.16200389 - 4.02276069i$	0.813sec	344
17			0.969sec	408
18	$+0.33265868 + 0.30837019i$	$+5.74711504 + 5.32749941i$	1.162sec	489
19			1.360sec	570
20	$+0.72087107 + 0.12470835i$	$+14.57220108 + 2.52094345i$	1.600sec	670
21			1.845sec	770
22	$+0.12317675 + 0.71091716i$	$+2.87061477 + 16.56781205i$	2.137sec	891
23			2.435sec	1012
24	$+0.33057618 + 0.33089982i$	$+8.77333085 + 8.78192027i$	2.781sec	1156

TS-code:
dabcgefjhimklon
jnmedclgoabhfki
lghmanbcfojedik

Table 42: Values of the quantum invariants for the 3-manifold induced by $5_2(-1)$

$$R_{30}(19) \in C_{30}(2)$$
TS-class with 2 rigid 3-gems
homology sphere, $\pi_1 = \pi_1' = \langle 11, 3, 2 \rangle$

r	$S^1 \times S^2$ – normalization	S^3 – normalization	$C^{++}, 486$ time	r – states
3	$+0.70710678$	$+1.00000000$	$0.008sec$	2
4	$+0.50000000$	$+1.00000000$	$0.014sec$	6
5	$-0.30075048 - 0.21850801i$	$-0.80901699 - 0.58778525i$	$0.025sec$	10
6	$+0.28867513$	$+1.00000000$	$0.047sec$	19
7	$-0.49248106 - 0.32673246i$	$-2.12348980 - 1.40881165i$	$0.070sec$	28
8	-0.73253782	-3.82842712	$0.110sec$	44
9	$+0.10190726 + 0.87979479i$	$+0.63206204 + 5.45677388i$	$0.151sec$	60
10	$+0.04270510 - 0.13143278i$	$+0.30901699 - 0.95105652i$	$0.214sec$	85
11	$+0.14806345 - 0.09583956i$	$+1.23251486 - 0.79779095i$	$0.280sec$	110
12	$+0.10566243$	$+1.00000000$	$0.372sec$	146
13	$-0.59773716 - 0.72632042i$	$-6.36789375 - 7.73773415i$	$0.467sec$	182
14	$-0.44911821 + 0.51624038i$	$-5.33996994 + 6.13804566i$	$0.593sec$	231
15	$+0.03961882 + 0.51996924i$	$+0.52185906 + 6.84903487i$	$0.723sec$	280
16	$+0.09754516 - 0.34675996i$	$+1.41421356 - 5.02733949i$	$0.888sec$	344
17	$+0.22450877 + 0.22923964i$	$+3.56218581 + 3.63724847i$	$1.059sec$	408
18	$-0.19211727 - 0.53326852i$	$-3.31907786 - 9.21291311i$	$1.270sec$	489
19	$-0.63606933 - 0.31120941i$	$-11.91106787 - 5.82772389i$	$1.486sec$	570
20	$-0.19227088 + 0.87403537i$	$-3.88670054 + 17.66837334i$	$1.747sec$	670
21	$+0.08967866 - 0.10170413i$	$+1.94972927 - 2.21117841i$	$2.014sec$	770
22	$+0.17608435 - 0.05170288i$	$+4.10361785 - 1.20492733i$	$2.334sec$	891
23	$+0.11564035 + 0.02683285i$	$+2.87996747 + 0.66825933i$	$2.657sec$	1012

The recognition of $\langle 11, 3, 2 \rangle = \langle a, b, c \mid a^{11} = b^3 = c^2 = abc \rangle$ as the fundamental group of $R_{30}(19)$ is due to Said Sidki (personal communication).

TS-code:

dabcgefihkjmlon
ikmedljgahoncfb
lifjmkbagnchode

Table 43: Values of the quantum invariants for the 3-manifold induced by $5_2(0)$

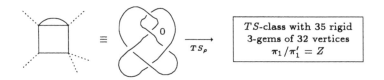

r	$S^1 \times S^2$ – *normalization*	S^3 – *normalization*	$C^{++}, 486$ *time*	$r-$ *states*
3	$+1.00000000$	$+1.41421356$	$0.008sec$	2
4	$+1.00000000$	$+2.00000000$	$0.016sec$	6
5	$+0.19098301 - 0.58778525i$	$+0.51374315 - 1.58113883i$	$0.027sec$	10
6			$0.051sec$	19
7	$-0.40096887 - 0.19309643i$	$-1.72890569 - 0.83259710i$	$0.077sec$	28
8	-0.41421356	-2.16478440	$0.119sec$	44
9	$-1.11334080 + 0.64278761i$	$-6.90530231 + 3.98677815i$	$0.165sec$	60
10	$+1.17557050i$	$+8.50650808i$	$0.234sec$	85
11	$+0.58458499 + 0.90963200i$	$+4.86622254 + 7.57199007i$	$0.306sec$	110
12	$+0.63397460 - 0.36602540i$	$+6.00000000 - 3.46410162i$	$0.405sec$	146
13	$-0.01444254 - 0.11894503i$	$-0.15386122 - 1.26716115i$	$0.509sec$	182
14	$-0.34601074 - 0.43388374i$	$-4.11403252 - 5.15883360i$	$0.647sec$	231
15	$-0.47814760 - 0.15535957i$	$-6.29816019 - 2.04639630i$	$0.788sec$	280
16	$-1.30656296 + 0.30656296i$	$-18.94260101 + 4.44456186i$	$0.967sec$	344
17	$-0.29971627 + 1.18671811i$	$-4.75547133 + 18.82915478i$	$1.154sec$	408
18	$+0.40760373 + 1.02606043i$	$+7.04188907 + 17.72653955i$	$1.383sec$	489
19	$+0.73355086 - 0.27049911i$	$+13.73651226 - 5.06538059i$	$1.619sec$	570
20	$-0.17557050i$	$-3.54910719i$	$1.902sec$	670
21	$-0.31195685 - 0.51659596i$	$-6.78234267 - 11.23145980i$	$2.193sec$	770
22	$-0.49611427 - 0.21510876i$	$-11.56186450 - 5.01307555i$	$2.540sec$	891
23	$-1.35284661 + 0.16912429i$	$-33.69199501 + 4.21195920i$	$2.892sec$	1012
24	$-0.41773767 + 1.17303261i$	$-11.08655439 + 31.13171449i$	$3.299sec$	1156
25	$+0.33015306 + 1.06269751i$	$+9.31331065 + 29.97770811i$	$3.727sec$	1300

TS-code:
$cabfdeighljknmpo$
$ipncmkfoabgeldhj$
$omgpclebniafhjkd$

Table 44: Values of the quantum invariants for the 3-manifold induced by $5_2(1)$

| | | | TS-class with 250 rigid 3-gems of 34 vertices $\pi_1 = \pi_1'$ homology sphere |

r	$S^1 \times S^2$ – normalization	S^3 – normalization	$C^{++}, 486$ time	$r-$ states
3	$+0.70710678$	$+1.00000000$	$0.008sec$	2
4	$+0.50000000$	$+1.00000000$	$0.017sec$	6
5	$-0.04387890 - 0.57206140i$	$-0.11803399 - 1.53884177i$	$0.030sec$	10
6	$+0.28867513$	$+1.00000000$	$0.055sec$	19
7	$+0.51544842 - 0.22610588i$	$+2.22252093 - 0.97492791i$	$0.083sec$	28
8	$+0.57402515$	$+3.00000000$	$0.129sec$	44
9	$-0.19777768 - 0.52483389i$	$-1.22668160 - 3.25519073i$	$0.178sec$	60
10	$-0.32532889 - 0.05020285i$	$-2.35410197 - 0.36327126i$	$0.253sec$	85
11	$-0.17361161 - 0.35370761i$	$-1.44518373 - 2.94434513i$	$0.329sec$	110
12	$+0.42264973 - 0.18301270i$	$+4.00000000 - 1.73205081i$	$0.437sec$	146
13	$-0.08469368 - 0.54601809i$	$-0.90227009 - 5.81691324i$	$0.548sec$	182
14	$-0.34629808 + 0.15676499i$	$-4.11744901 + 1.86391983i$	$0.696sec$	231
15	$-0.70159629 + 0.17629311i$	$-9.24142629 + 2.32213284i$	$0.848sec$	280
16	$-0.06517759 + 0.22632747i$	$-0.94494729 + 3.28130457i$	$1.042sec$	344
17	$-0.26332845 - 0.37366224i$	$-4.17812119 - 5.92874081i$	$1.241sec$	408
18	$-0.03451354 + 0.10025582i$	$-0.59626666 + 1.73205081i$	$1.489sec$	489
19	$-0.32392608 + 0.51257414i$	$-6.06585686 + 9.59849037i$	$1.741sec$	570
20	$+0.31518075 + 0.80125652i$	$+6.37128803 + 16.19716980i$	$2.047sec$	670
21	$-0.06818995 + 0.15610051i$	$-1.48253701 + 3.39382573i$	$2.360sec$	770
22	$+0.12832880 - 0.07274939i$	$+2.99068249 - 1.69541312i$	$2.732sec$	891
23	$-0.13894617 + 0.06668791i$	$-3.46038765 + 1.66083029i$	$3.111sec$	1012
24	$+0.77176762 + 0.34130829i$	$+20.48233707 + 9.05815583i$	$3.554sec$	1156
25	$+0.48976598 + 0.12317374i$	$+13.81584267 + 3.47461647i$	$4.007sec$	1300

TS-code:
| $dabcgefjhimklonqp$ |
| $jipodlqgbacfnmekh$ |
| $qgnjpkocfmiedhbla$ |

14.6 Knot 6_1

Table 45: Values of the quantum invariants for the 3-manifold induced by $6_1(-2)$

6 rigid 3-gems with 34 vertices
$$\pi_1/\pi_1' = Z_2$$
$$\pi_1'/\pi_1'' = Z_9$$
$$\pi_1''/\pi_1''' = Z_2^6 \oplus Z_4^2$$

r	$S^1 \times S^2$ – normalization	S^3 – normalization	$C^{++}, 486$ time	$r-$ states
3			0.006sec	2
4	-0.38268343	-0.76536686	0.010sec	6
5			0.018sec	10
6	$-0.28867513 - 0.86602540i$	$-1.00000000 - 3.00000000i$	0.035sec	19
7			0.052sec	28
8	$+0.19509032 + 0.28726536i$	$+1.01959116 + 1.50132111i$	0.081sec	44
9			0.113sec	60
10	$-0.23860231 + 0.85065081i$	$-1.72654253 + 6.15536707i$	0.160sec	85
11			0.210sec	110
12	$+0.06820648 - 0.50971567i$	$+0.64551307 - 4.82400088i$	0.279sec	146
13			0.351sec	182
14	$+0.63596804 - 0.44896104i$	$+7.56159543 - 5.33810121i$	0.445sec	231
15			0.544sec	280
16	$-0.49113112 + 0.39778375i$	$-7.12043814 + 5.76708441i$	0.669sec	344
17			0.799sec	408
18	$-0.64086712 + 0.04054948i$	$-11.07181989 + 0.70054546i$	0.958sec	489
19			1.123sec	570
20	$+0.72029317 - 0.09351838i$	$+14.56051895 - 1.89044715i$	1.320sec	670
21			1.525sec	770
22	$+0.45969896 + 0.29575466i$	$+10.71321159 + 6.89251547i$	1.765sec	891
23			2.013sec	1012

TS-code:
$$dabcgefjhimklonqp$$
$$jomedlhgknqfcapbi$$
$$qihnpkaobfjedmclg$$

Table 46: Values of the quantum invariants for the 3-manifold induced by $6_1(-1)$

r	$S^1 \times S^2$ $-$ *normalization*	S^3 $-$ *normalization*	$C^{++}, 486$ *time*	$r-$ *states*
3	$+0.70710678$	$+1.00000000$	$0.007sec$	2
4	$+0.50000000$	$+1.00000000$	$0.013sec$	6
5	$-0.04387890 - 0.57206140i$	$-0.11803399 - 1.53884177i$	$0.024sec$	10
6	$+0.28867513$	$+1.00000000$	$0.044sec$	19
7	$+0.51544842 - 0.22610588i$	$+2.22252093 - 0.97492791i$	$0.068sec$	28
8	$+0.57402515$	$+3.00000000$	$0.107sec$	44
9	$-0.19777768 - 0.52483389i$	$-1.22668160 - 3.25519073i$	$0.150sec$	60
10	$-0.32532889 - 0.05020285i$	$-2.35410197 - 0.36327126i$	$0.212sec$	85
11	$-0.17361161 - 0.35370761i$	$-1.44518373 - 2.94434513i$	$0.279sec$	110
12	$+0.42264973 - 0.18301270i$	$+4.00000000 - 1.73205081i$	$0.370sec$	146
13	$-0.08469368 - 0.54601809i$	$-0.90227009 - 5.81691324i$	$0.468sec$	182
14	$-0.34629808 + 0.15676499i$	$-4.11744901 + 1.86391983i$	$0.594sec$	231
15	$-0.70159629 + 0.17629311i$	$-9.24142629 + 2.32213284i$	$0.727sec$	280
16	$-0.06517759 + 0.22632747i$	$-0.94494729 + 3.28130457i$	$0.894sec$	344
17	$-0.26332845 - 0.37366224i$	$-4.17812119 - 5.92874081i$	$1.069sec$	408
18	$-0.03451354 + 0.10025582i$	$-0.59626666 + 1.73205081i$	$1.281sec$	489
19	$-0.32392608 + 0.51257414i$	$-6.06585686 + 9.59849037i$	$1.505sec$	570
20	$+0.31518075 + 0.80125652i$	$+6.37128803 + 16.19716980i$	$1.768sec$	670
21	$-0.06818995 + 0.15610051i$	$-1.48253701 + 3.39382573i$	$2.045sec$	770
22	$+0.12832880 - 0.07274939i$	$+2.99068249 - 1.69541312i$	$2.366sec$	891
23	$-0.13894617 + 0.06668791i$	$-3.46038765 + 1.66083029i$	$2.701sec$	1012
24	$+0.77176762 + 0.34130829i$	$+20.48233707 + 9.05815583i$	$3.086sec$	1156
25	$+0.48976598 + 0.12317374i$	$+13.81584267 + 3.47461647i$	$3.484sec$	1300

TS-code:
> $dabcgefjhimklonqp$
> $jipodlqgbacfnmekh$
> $qgnjpkocfmiedhbla$

Table 47: Values of the quantum invariants for the 3-manifold induced by $6_1(0)$

r	$S^1 \times S^2 - normalization$	$S^3 - normalization$	$C^{++}, 486$ time	$r-$ states
3	$+1.00000000$	$+1.41421356$	$0.008sec$	2
4	$+1.00000000$	$+2.00000000$	$0.015sec$	6
5	$+0.69098301 - 0.95105652i$	$+1.85874017 - 2.55833637i$	$0.026sec$	10
6	$+0.50000000 - 0.86602540i$	$+1.73205081 - 3.00000000i$	$0.050sec$	19
7	$+0.37651020 - 0.78183148i$	$+1.62344430 - 3.37111682i$	$0.076sec$	28
8	-0.41421356	-2.16478440	$0.119sec$	44
9	$-0.26604444 + 0.22323779i$	$-1.65009430 + 1.38459352i$	$0.167sec$	60
10	$-0.11803399 + 0.36327126i$	$-0.85410197 + 2.62865556i$	$0.236sec$	85
11	$+0.97592460 + 0.73091318i$	$+8.12382528 + 6.08429275i$	$0.311sec$	110
12	$+1.13397460 + 0.50000000i$	$+10.73205081 + 4.73205081i$	$0.411sec$	146
13	$+1.20602247 + 0.28867004i$	$+12.84816048 + 3.07529839i$	$0.520sec$	182
14	$+1.00000000 - 0.86776748i$	$+11.88989846 - 10.31766721i$	$0.660sec$	231
15	$+0.70448853 - 0.90949078i$	$+9.27952292 - 11.97981257i$	$0.808sec$	280
16	$+0.45880390 - 0.87301746i$	$+6.65175690 - 12.65704134i$	$0.992sec$	344
17	$-0.47332529 - 0.11683212i$	$-7.51005235 - 1.85372590i$	$1.187sec$	408
18	$-0.37938524 + 0.18198512i$	$-6.55437759 + 3.14403156i$	$1.422sec$	489
19	$-0.22902045 + 0.40149936i$	$-4.28864905 + 7.51849821i$	$1.669sec$	570
20	$+0.89680225 + 0.85796048i$	$+18.12859915 + 17.34342399i$	$1.962sec$	670
21	$+1.12412804 + 0.63611857i$	$+24.43998777 + 13.83003506i$	$2.267sec$	770
22	$+1.25309761 + 0.39382757i$	$+29.20324159 + 9.17808923i$	$2.624sec$	891
23	$+1.09831068 - 0.81992663i$	$+27.35289997 - 20.41987891i$	$2.994sec$	1012
24	$+0.78936911 - 0.91773767i$	$+20.94947209 - 24.35631104i$	$3.419sec$	1156
25	$+0.50445311 - 0.91547404i$	$+14.23015300 - 25.82467079i$	$3.863sec$	1300

TS-code:

```
dabcgefjhilknmpo
gplndiobfmjckeah
ikempblganfhdojc
```

Table 48: Values of the quantum invariants for the 3-manifold induced by $6_1(1)$

r	$S^1 \times S^2$ – *normalization*	S^3 – *normalization*	$C^{++}, 486$ *time*	$r-$ *states*
3	+0.70710678	+1.00000000	0.008*sec*	2
4	+0.50000000	+1.00000000	0.016*sec*	6
5	+0.37174803 − 0.70710678i	+1.00000000 − 1.90211303i	0.029*sec*	10
6	+0.28867513	+1.00000000	0.055*sec*	19
7	−0.05160719 + 0.22610588i	−0.22252093 + 0.97492791i	0.083*sec*	28
8	−0.73253782	−3.82842712	0.130*sec*	44
9	−0.17786213 + 0.40378134i	−1.10315889 + 2.50438339i	0.182*sec*	60
10	−0.36180340 + 0.52573111i	−2.61803399 + 3.80422607i	0.257*sec*	85
11	−0.75817761 + 0.51175031i	−6.31124829 + 4.25992960i	0.339*sec*	110
12	+0.10566243	+1.00000000	0.449*sec*	146
13	+0.10781436 + 0.05658534i	+1.14858242 + 0.60282255i	0.567*sec*	182
14	+0.05243861 + 0.06575594i	+0.62348980 + 0.78183148i	0.719*sec*	231
15	+0.39786225 − 0.87662282i	+5.24064161 − 11.54687577i	0.880*sec*	280
16	+0.49039264 − 0.48470965i	+7.10973169 − 7.02733949i	1.081*sec*	344
17	+0.40628399 − 0.27459952i	+6.44633642 − 4.35695450i	1.292*sec*	408
18	−0.05909519 − 0.76366981i	−1.02094684 − 13.19339741i	1.549*sec*	489
19	+0.26144886 − 0.11240560i	+4.89590510 − 2.10491313i	1.816*sec*	570
20	+0.06233884 + 0.24986877i	+1.26016163 + 5.05102526i	2.135*sec*	670
21	−0.65901113 + 0.33123996i	−14.32774870 + 7.20158229i	2.466*sec*	770
22	+0.08147124 + 0.47875344i	+1.89867433 + 11.15727309i	2.854*sec*	891

TS-code:	$dabcgefjhilknmpo$
	$jmoedchgpniblafk$
	$pnikojaecmdhfblg$

14.7 Knot 6_2

Table 49: Values of the quantum invariants for the 3-manifold induced by $6_2(-1)$

r	$S^1 \times S^2$ − normalization	S^3 − normalization	$C^{++}, 486$ time	$r-$ states
3	$+0.70710678$	$+1.00000000$	$0.008sec$	2
4	-0.50000000	-1.00000000	$0.015sec$	6
5	$-0.04387890 - 0.13504538i$	$-0.11803399 - 0.36327126i$	$0.025sec$	10
6	$+0.28867513$	$+1.00000000$	$0.048sec$	19
7	$-0.05160719 + 0.06471337i$	$-0.22252093 + 0.27903243i$	$0.072sec$	28
8	$+0.57402515i$	$+3.00000000i$	$0.113sec$	44
9	$+0.12053964 + 0.51002386i$	$+0.74762613 + 3.16333415i$	$0.156sec$	60
10	$-0.01631190 - 0.01185129i$	$-0.11803399 - 0.08575671i$	$0.222sec$	85
11	$-0.66084533 - 0.20657019i$	$-5.50103155 - 1.71953873i$	$0.290sec$	110
12	$-0.28867513 + 0.04903811i$	$-2.73205081 + 0.46410162i$	$0.384sec$	146
13	$+0.05941008 + 0.90647311i$	$+0.63291542 + 9.65696102i$	$0.483sec$	182
14	$-0.66606008 + 0.32075763i$	$-7.91938675 + 3.81377565i$	$0.613sec$	231
15	$-0.24965598 - 0.37914823i$	$-3.28846850 - 4.99414043i$	$0.748sec$	280
16	$+0.29767452 + 0.25726999i$	$+4.31569680 + 3.72991036i$	$0.920sec$	344
17	$+0.38191156 - 0.02031338i$	$+6.05962937 - 0.32230376i$	$1.096sec$	408
18	$+0.30424575 - 0.49069209i$	$+5.25624426 - 8.47734945i$	$1.314sec$	489
19	$-0.27331010 - 0.42466823i$	$-5.11801944 - 7.95235979i$	$1.539sec$	570
20	$-0.07714048 - 0.16462140i$	$-1.55937261 - 3.32777427i$	$1.809sec$	670
21	$+0.42796021 - 0.11461109i$	$+9.30440485 - 2.49179228i$	$2.086sec$	770
22	$+0.71847941 - 0.27315226i$	$+16.74404900 - 6.36577009i$	$2.417sec$	891
23	$+0.27739236 + 0.33882372i$	$+6.90832347 + 8.43824212i$	$2.753sec$	1012

TS-code:
dabcgefjhimklpnorqtsvu
jkprdvsgnlbaimtcoehuqf
niuftocqbpjesadkvmrlgh

Table 50: Values of the quantum invariants for the 3-manifold induced by $6_2(0)$

r	$S^1 \times S^2$ – normalization	S^3 – normalization	C^{++}, 486 time	$r-$ states
3	+1.00000000	+1.41421356	0.007sec	2
4			0.013sec	6
5	+0.50000000 − 0.36327126i	+1.34499702 − 0.97719754i	0.023sec	10
6	+1.00000000	+3.46410162	0.043sec	19
7	+0.80193774	+3.45781137	0.065sec	28
8	+0.29289322 + 0.29289322i	+1.53073373 + 1.53073373i	0.102sec	44
9	−0.15270364 + 0.41954982i	−0.94711775 + 2.60218463i	0.141sec	60
10	−0.19098301 + 0.58778525i	−1.38196601 + 4.25325404i	0.200sec	85
11	−0.07773241 + 0.28494055i	−0.64706285 + 2.37191193i	0.261sec	110
12	−0.36602540 + 0.36602540i	−3.46410162 + 3.46410162i	0.346sec	146
13	−1.29949059 − 0.64001808i	−13.84390757 − 6.81832650i	0.436sec	182
14	−0.33243720 − 0.97492791i	−3.95264453 − 11.59179389i	0.554sec	231
15	−0.46805576 − 0.12430007i	−6.16523040 − 1.63728047i	0.676sec	280
16	−0.23784051 − 0.94173992i	−3.44822103 − 13.65338217i	0.831sec	344
17	+0.60950581 − 1.00917042i	+9.67077117 − 16.01208060i	0.991sec	408
18	+0.68479253 − 0.86602540i	+11.83068904 − 14.96172460i	1.188sec	489
19	+0.55704936 − 0.28303887i	+10.43133569 − 5.30020084i	1.391sec	570
20	+0.34843503 + 0.12224865i	+7.04351376 + 2.47122130i	1.636sec	670
21	+0.44225153 + 0.46607333i	+9.61511644 + 10.13303289i	1.887sec	770
22	+0.48246238 + 0.33958282i	+11.24370935 + 7.91392388i	2.186sec	891
23	+0.13268291 + 0.69891859i	+3.30440410 + 17.40623184i	2.490sec	1012
24	−1.14480464 + 0.03175208i	−30.38255791 + 0.84268462i	2.845sec	1156
25	−0.43087070 − 0.41103525i	−12.15446163 − 11.59492183i	3.211sec	1300

TS-code:
```
dabcgefjhimklpnosqrut
junrdmlgkaqhfstipecob
oslpitamunjcrkgdfhbqe
```

Table 51: Values of the quantum invariants for the 3-manifold induced by $6_2(1)$

r	$S^1 \times S^2 - normalization$	$S^3 - normalization$	$C^{++}, 486$ time	$r-$ states
3	+0.70710678	+1.00000000	0.007sec	2
4	−0.50000000	−1.00000000	0.012sec	6
5	+0.11487646 − 0.35355339i	+0.30901699 − 0.95105652i	0.021sec	10
6	+0.28867513	+1.00000000	0.040sec	19
7	+0.30194620 − 0.23497571i	+1.30193774 − 1.01317303i	0.060sec	28
8	−0.34985438i	−1.82842712i	0.094sec	44
9	+0.04406708 − 0.32190203i	+0.27331840 − 1.99654128i	0.130sec	60
10	−0.11180340 + 0.08122992i	−0.80901699 + 0.58778525i	0.184sec	85
11	−0.43110476 + 0.45761139i	−3.58861716 + 3.80926453i	0.241sec	110
12	−0.28867513 − 0.31698730i	−2.73205081 − 3.00000000i	0.321sec	146
13	+0.34090107 + 0.42567195i	+3.63173308 + 4.53482551i	0.404sec	182
14	−0.64414626 − 0.47172650i	−7.65883360 − 5.60878016i	0.512sec	231
15	+0.39271888 − 0.62110535i	+5.17289314 − 8.18119964i	0.626sec	280
16	−0.40823194 − 0.36183156i	−5.91856261 − 5.24584806i	0.769sec	344
17	−0.17575336 − 0.52515534i	−2.78860424 − 8.33241789i	0.917sec	408
18	−0.08924874 − 0.35089538i	−1.54188907 − 6.06217783i	1.102sec	489
19	+0.26452295 + 0.10237743i	+4.95347072 + 1.91712515i	1.290sec	570
20	+0.12075728 + 0.08398846i	+2.44107374 + 1.69780247i	1.517sec	670
21	+0.27832657 − 0.16664442i	+6.05117713 − 3.62306370i	1.750sec	770
22	+0.51151841 − 0.04215215i	+11.92085573 − 0.98234922i	2.028sec	891
23	−0.54070205 − 0.08365131i	−13.46592486 − 2.08329573i	2.311sec	1012
24	+0.14074617 − 0.79501879i	+3.73533475 − 21.09941163i	2.642sec	1156
25	−0.44179394 + 0.34109184i	−12.46259590 + 9.62188339i	2.979sec	1300
26	−0.35537178 − 0.64126639i	−10.63005182 − 19.18186929i	3.371sec	1469
27	+0.20415721 + 0.00305900i	+6.46138893 + 0.09681451i	3.768sec	1638

Table 52: Values of the quantum invariants for the 3-manifold induced by $6_2(2)$

r	$S^1 \times S^2$ − normalization	S^3 − normalization	$C^{++}, 486$ time	$r-$ states
3			0.006sec	2
4	$+0.92387953i$	$+1.84775907i$	0.011sec	6
5			0.021sec	12
6	$+0.21132487$	$+0.73205081$	0.040sec	23
7			0.066sec	40
8	$-0.57845756 - 0.25301205i$	$-3.02316490 - 1.32230471i$	0.107sec	66
9			0.161sec	102
10	$+0.39143445 - 0.41332264i$	$+2.83244626 - 2.99083070i$	0.238sec	153
11			0.335sec	220
12	$-0.21257839 + 1.08562058i$	$-2.01186345 + 10.27442351i$	0.463sec	308
13			0.622sec	420
14	$+0.35145526 - 0.69034442i$	$+4.17876733 - 8.20812509i$	0.821sec	561
15			1.060sec	734
16	$-0.40795723 - 0.46059424i$	$-5.91457994 - 6.67771336i$	1.353sec	946
17			1.699sec	1200
18	$+0.28726390 + 0.98025656i$	$+4.96286058 + 16.93521768i$	2.111sec	1503
19			2.591sec	1860
20	$-0.43474791 - 0.42755126i$	$-8.78830380 - 8.64282562i$	3.152sec	2278
21			3.794sec	2762
22	$+0.73406345 - 0.12781307i$	$+17.10723255 - 2.97866344i$	4.537sec	3321
23			5.379sec	3960
24	$-0.37758573 - 0.11336880i$	$-10.02094160 - 3.00875282i$	6.336sec	4688
25			7.418sec	5512

TS-code:

> $dabcgefjhimklpnorqts$
> $gqpldkamofjesbrtnihc$
> $kmfotqhgbpanildrcsje$

14.8   Knot 6_3

Table 53: Values of the quantum invariants for the 3-manifold induced by $6_3(0)$

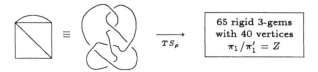

r	$S^1 \times S^2$ − normalization	S^3 − normalization	$C^{++}, 486$ time	$r-$ states
3	+1.00000000	+1.41421356	0.007sec	2
4			0.019sec	10
5	+0.38196601	+1.02748630	0.049sec	26
6	+1.00000000	+3.46410162	0.120sec	65
7	+0.19806226	+0.85400888	0.260sec	152
8	−0.41421356	−2.16478440	0.516sec	312
9	−0.22668160	−1.40595311	0.950sec	606
10	−0.23606798	−1.70820393	1.671sec	1101
11	−0.01138665	−0.09478511	2.786sec	1906
12	+0.26794919	+2.53589838	4.477sec	3144
13	−0.18693624	−1.99149419	6.977sec	5020
14	+0.06099892	+0.72527102	10.546sec	7727
15	+1.27294542	+16.76723700	15.526sec	11588
16	+1.05505271	+15.29619546	22.383sec	16930
17	−0.03065179	−0.48633901	31.649sec	24212
18	+0.40373334	+6.97502295	43.928sec	33915
19	+0.99528209	+18.63770498	60.008sec	46724
20	+0.17015854	+3.43970595	80.695sec	63292
21	−0.41712284	−9.06878653	107.269sec	84552
22	−0.23573339	−5.49372924	140.808sec	111437
23	−0.24902554	−6.20186151	182.719sec	145154

TS-code:
$$\begin{array}{l} dabchefgkijnlmporqts \\ kslednpihgacrfmtboqj \\ tgqmljbanreksoicpfdh \end{array}$$

Table 54: Values of the quantum invariants for the 3-manifold induced by $6_3(-1)$

r	$S^1 \times S^2$ − normalization	S^3 − normalization	$C^{++}, 486$ time	$r-$ states
3	+0.70710678	+1.00000000	0.007 sec	2
4	−0.50000000	−1.00000000	0.021 sec	10
5	−0.04387890 + 0.13504538i	−0.11803399 + 0.36327126i	0.074 sec	42
6	+0.28867513	+1.00000000	0.206 sec	123
7	−0.20895325 + 0.62466477i	−0.90096887 + 2.69344219i	0.522 sec	338
8	+0.73253782i	+3.82842712i	1.188 sec	800
9	+0.09888884 + 0.23458321i	+0.61334080 + 1.45496147i	2.521 sec	1760
10	−0.01631190 + 0.01185129i	−0.11803399 + 0.08575671i	4.996 sec	3553

Table 55: Values of the quantum invariants for the 3-manifold induced by $6_3(1)$

r	$S^1 \times S^2$ − normalization	S^3 − normalization	$C^{++}, 486$ time	$r-$ states
3	+0.70710678	+1.00000000	0.007 sec	2
4	−0.50000000	−1.00000000	0.021 sec	10
5	−0.04387890 − 0.13504538i	−0.11803399 − 0.36327126i	0.074 sec	42
6	+0.28867513	+1.00000000	0.205 sec	123

TS-codes of both blinks in this page:

$$dabchefgkijnlmporqts$$
$$kslednpihgacrfmtboqj$$
$$tgqmljbanreksoicpfdh$$

14.9   Knot 7_1

Table 56: Values of the quantum invariants for the 3-manifold induced by $7_1(7)$

r	$S^1 \times S^2$ − *normalization*	S^3 − *normalization*	$C^{++}, 486$ *time*	$r-$ *states*
3	$+0.70710678i$	$+1.00000000i$	$0.006sec$	2
4	$+0.35355339 + 0.35355339i$	$+0.70710678 + 0.70710678i$	$0.008sec$	4
5	$-0.92561479 - 0.07099756i$	$-2.48989828 - 0.19098301i$	$0.011sec$	6
6	$+0.28867513$	$+1.00000000$	$0.017sec$	9
7			$0.022sec$	12
8	$-0.07322330 - 0.17677670i$	$-0.38268343 - 0.92387953i$	$0.030sec$	16
9	$+1.06650911 + 0.00972335i$	$+6.61483690 + 0.06030738i$	$0.037sec$	20
10	$-0.85172209 + 0.13143278i$	$-6.16311896 + 0.95105652i$	$0.046sec$	25
11	$-0.68142338 - 0.08245717i$	$-5.67232803 - 0.68639280i$	$0.056sec$	30
12	$+0.98594555 - 0.27883877i$	$+9.33108886 - 2.63895843i$	$0.067sec$	36
13	$-0.30182402 + 0.18103838i$	$-3.21543215 + 1.92866234i$	$0.078sec$	42
14			$0.090sec$	49
15	$+0.13192125 + 0.25749264i$	$+1.73766672 + 3.39169306i$	$0.103sec$	56
16	$-1.05650193 - 0.28405437i$	$-15.31720631 - 4.11823144i$	$0.118sec$	64
17	$+0.88733185 + 0.05423876i$	$+14.07891954 + 0.86058341i$	$0.133sec$	72
18	$+0.72519446 + 0.37152172i$	$+12.52868532 + 6.41852487i$	$0.149sec$	81
19	$-1.06577663 - 0.18150753i$	$-19.95778949 - 3.39891964i$	$0.166sec$	90
20	$+0.33010783 - 0.11539543i$	$+6.67303478 - 2.33268536i$	$0.184sec$	100
21			$0.202sec$	110
22	$-0.05809091 - 0.32596736i$	$-1.35379943 - 7.59661786i$	$0.223sec$	121
23	$+0.74294138 + 0.78809535i$	$+18.50259813 + 19.62713607i$	$0.243sec$	132
24	$-0.71158817 - 0.44119608i$	$-18.88520364 - 11.70912922i$	$0.265sec$	144

TS-code:

$dabchefgkijnlmqopsrut$
$hpltduranfckosmbgqiej$
$nigfodcbhamsujrtleqpk$

Table 57: Values of the quantum invariants for the 3-manifold induced by $7_1(-1)$

a single rigid 3-gem
with 60 vertices
$\pi_1 = \pi_1'$
homology sphere

r	$S^1 \times S^2$ — normalization	S^3 — normalization	C^{++}, 486 time	$r-$ states
3	+0.70710678	+1.00000000	0.010*sec*	2
4	+0.50000000	+1.00000000	0.015*sec*	4
5	−0.30075048 + 0.92561479i	−0.80901699 + 2.48989828i	0.023*sec*	6
6	+0.28867513	+1.00000000	0.035*sec*	9
7	+0.14460014 − 0.18132284i	+0.62348980 − 0.78183148i	0.046*sec*	12
8	−0.19134172	−1.00000000	0.061*sec*	16
9	−0.81015012 − 0.35355339i	−5.02481493 − 2.19285330i	0.073*sec*	20
10	−0.76631190 + 0.55675818i	−5.54508497 + 4.02874005i	0.092*sec*	25
11	+0.36686185 + 0.25402482i	+3.05384412 + 2.11456219i	0.110*sec*	30
12	+0.78867513 − 0.18301270i	+7.46410162 − 1.73205081i	0.132*sec*	36
13	+0.42149852 + 0.60111287i	+4.49036462 + 6.40385629i	0.154*sec*	42
14	−0.07577599 + 0.03649180i	−0.90096887 + 0.43388374i	0.179*sec*	49
15	−0.57474573 − 0.41757721i	−7.57055071 − 5.50032705i	0.204*sec*	56
16	−0.06897484i	−1.00000000i	0.233*sec*	64
17	+0.44249150 − 0.19014015i	+7.02082556 − 3.01687346i	0.263*sec*	72
18	−0.36576545 + 0.43301270i	−6.31907786 + 7.48086230i	0.295*sec*	81
19	−0.41061473 + 0.64375142i	−7.68919308 + 12.05492311i	0.328*sec*	90
20	+0.17264150 − 0.35235568i	+3.48989828 − 7.12276851i	0.364*sec*	100
21	+0.27815666 + 0.02323028i	+6.04748324 + 0.50505606i	0.401*sec*	110
22	+0.16610378 + 0.39538302i	+3.87102225 + 9.21433874i	0.441*sec*	121
23	+0.02938276 − 0.34991283i	+0.73176346 − 8.71441107i	0.480*sec*	132
24	+0.14376734 + 0.58487348i	+3.81551515 + 15.52225745i	0.524*sec*	144
25	+0.13823483 + 0.68798420i	+3.89947598 + 19.40739412i	0.568*sec*	156

TS-code ($A = 27, B = 28, \ldots$):

$habcdefgpijklmnorqtsvuxwzyBADC$
$rwAtlkjihgfesBxqpvCdoazunbDymc$
$onmlqxBstAwrcbahudpDekvyfjzCgi$

Table 58: Values of the quantum invariants for the 3-manifold induced by $7_1(0)$

r	$S^1 \times S^2 - normalization$	$S^3 - normalization$	$C^{++}, 486$ time	$r-$ states
3	+1.00000000	+1.41421356	0.010sec	2
4	+1.00000000	+2.00000000	0.014sec	4
5	−0.30901699 + 0.95105652i	−0.83125388 + 2.55833637i	0.021sec	6
6			0.031sec	9
7			0.041sec	12
8			0.055sec	16
9	+1.00000000	+6.20232575	0.069sec	20
10	−0.80901699 + 0.58778525i	−5.85410197 + 4.25325404i	0.086sec	25
11	−0.84125353 − 0.54064082i	−7.00279171 − 4.50042096i	0.103sec	30
12	+1.00000000	+9.46410162	0.123sec	36
13			0.151sec	42
14			0.175sec	49
15			0.194sec	56
16	+0.70710678 − 0.70710678i	+10.25166179 − 10.25166179i	0.221sec	64
17	−0.27366299 + 0.96182564i	−4.34209501 + 15.26088102i	0.246sec	72
18	−1.00000000	−17.27631145	0.277sec	81
19	+0.87947375 − 0.47594739i	+16.46907197 − 8.91261605i	0.308sec	90
20			0.342sec	100
21			0.375sec	110
22			0.413sec	121
23	+0.46006504 − 0.88788522i	+11.45769878 − 22.11235489i	0.451sec	132
24	+1.00000000i	+26.53951331i	0.491sec	144
25	−0.96858316 + 0.24868989i	−27.32282978 + 7.01531033i	0.532sec	156
26	+0.74851075 − 0.66312266i	+22.38981426 − 19.83564456i	0.577sec	169
27			0.621sec	182

TS-code $(A = 27, B = 28, \ldots)$:

habcde´fgoijklmnqpsrutwvyxAzCB
qrzwukjihgfevAsoaypdlBtnbCxmc
tnmlpsAvwzrdcbaekqxohuCfjyBgi

Table 59: Values of the quantum invariants for the 3-manifold induced by $7_1(1)$

r	$S^1 \times S^2 - normalization$	$S^3 - normalization$	$C^{++}, 486$ time	$r -$ states
3	+0.70710678	+1.00000000	0.009sec	2
4	+0.50000000	+1.00000000	0.013sec	4
5	−0.71637742 + 0.35355339i	−1.92705098 + 0.95105652i	0.020sec	6
6	+0.28867513	+1.00000000	0.029sec	9
7	+0.14460014 + 0.18132284i	+0.62348980 + 0.78183148i	0.039sec	12
8	−0.19134172	−1.00000000	0.052sec	16
9	+0.61574932 + 0.54167522i	+3.81907786 + 3.35964617i	0.064sec	20
10	+0.38819660 + 0.50655533i	+2.80901699 + 3.66546879i	0.080sec	25
11	−0.06825293 − 0.42302699i	−0.56815336 − 3.52137589i	0.096sec	30
12	+0.10566243	+1.00000000	0.115sec	36
13	−0.09113971 + 0.75060306i	−0.97094182 + 7.99642529i	0.135sec	42
14	−0.07577599 − 0.03649180i	−0.90096887 − 0.43388374i	0.157sec	49
15	+0.68040114 − 0.14125017i	+8.96224377 − 1.86054730i	0.179sec	56
16	−0.39284748 + 0.76249477i	−5.69551813 + 11.05467898i	0.205sec	64
17	−0.11842228 + 0.44553394i	−1.87895627 + 7.06909875i	0.230sec	72
18	+0.35408086 + 0.81379768i	+6.11721119 + 14.05942220i	0.259sec	81
19	+0.38764504 + 0.32299777i	+7.25906151 + 6.04847340i	0.288sec	90
20	−0.36614604 + 0.16561496i	−7.40153677 + 3.34785876i	0.319sec	100
21	−0.01023495 + 0.24435804i	−0.22252093 + 5.31265762i	0.351sec	110
22	+0.02302617 + 0.50499023i	+0.53662129 + 11.76871748i	0.386sec	121
23	+0.50925977 + 0.54785926i	+12.68286992 + 13.64417168i	0.422sec	132
24	−0.88230286 − 0.19134172i	−23.41588862 − 5.07811602i	0.460sec	144
25	+0.21092776 + 0.11136183i	+5.95007594 + 3.14141371i	0.498sec	156
26	−1.25169462 + 1.10890467i	−37.44129276 + 33.17009094i	0.540sec	169
27	−0.05808024 − 0.20576512i	−1.83818642 − 6.51227788i	0.581sec	182

TS-code $(A = 27, B = 28, \ldots)$:

> $gabcdefnhijklmporqtsvuxwzyBA$
> $puyrjihgfeqzvontAdmaxslbBwkc$
> $qmlkovzryupcbasdnBejtwfixAgh$

14.10 Knot 7_2

Table 60: Values of the quantum invariants for the 3-manifold induced by $7_2(-7)$

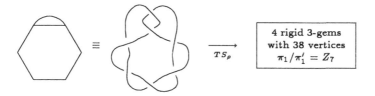

r	$S^1 \times S^2$ – normalization	S^3 – normalization	C^{++}, 486 time	$r-$ states
3	$-0.70710678i$	$-1.00000000i$	$0.007sec$	2
4	$-0.35355339 + 0.35355339i$	$-0.70710678 + 0.70710678i$	$0.012sec$	6
5	$+0.57206140 - 0.18587402i$	$+1.53884177 - 0.50000000i$	$0.020sec$	10
6	$+0.28867513$	$+1.00000000$	$0.038sec$	19
7	$+0.24603621 + 0.19620733i$	$+1.06086391 + 0.84601074i$	$0.058sec$	28
8	$+0.17677670 + 0.07322330i$	$+0.92387953 + 0.38268343i$	$0.090sec$	44
9	$+0.40204608 - 0.67691754i$	$+2.49362077 - 4.19846310i$	$0.125sec$	60
10	$-0.29270510 - 0.21266270i$	$-2.11803399 - 1.53884177i$	$0.178sec$	85
11	$+0.28438824 + 0.49512007i$	$+2.36731437 + 4.12149558i$	$0.232sec$	110
12	$+0.07471462 - 0.18410442i$	$+0.70710678 - 1.74238296i$	$0.308sec$	146
13	$+0.38232200 - 0.32964538i$	$+4.07300407 - 3.51182237i$	$0.388sec$	182
14	$+0.07742566 - 0.83922397i$	$+0.92058320 - 9.97828776i$	$0.493sec$	231
15	$-0.17108417 + 0.34229973i$	$-2.25352062 + 4.50877199i$	$0.602sec$	280
16	$+0.19896411 - 0.24342526i$	$+2.88458948 - 3.52918900i$	$0.739sec$	344
17	$-0.29275601 - 0.24825643i$	$-4.64503589 - 3.93897987i$	$0.882sec$	408
18	$+0.46472219 - 0.22113798i$	$+8.02868532 - 3.82044866i$	$1.059sec$	489
19	$+0.19176360 - 0.23829101i$	$+3.59097522 - 4.46225000i$	$1.240sec$	570
20	$-0.19366897 - 0.43897587i$	$-3.91496254 - 8.87377074i$	$1.457sec$	670
21	$-0.83085016 - 0.05597062i$	$-18.06374990 - 1.21687308i$	$1.682sec$	770
22	$+0.50438105 + 0.17169741i$	$+11.75452065 + 4.00138091i$	$1.948sec$	891
23	$+0.10682535 - 0.53905393i$	$+2.66043409 - 13.42487926i$	$2.220sec$	1012
24	$-0.30214653 + 0.08577901i$	$-8.01882174 + 2.27653305i$	$2.535sec$	1156

TS-code:

$dabcgefjhimklonqpsr$
$jsmedclgporhfiankqb$
$igsnhormakjdelpfcbq$

Table 61: Values of the quantum invariants for the 3-manifold induced by $7_2(-1)$

r	$S^1 \times S^2$ − normalization	S^3 − normalization	C^{++}, 486 time	$r-$ states
3	+0.70710678	+1.00000000	0.010sec	2
4	−0.50000000	−1.00000000	0.021sec	6
5	+0.37174803 + 0.70710678i	+1.00000000 + 1.90211303i	0.036sec	10
6	+0.28867513	+1.00000000	0.067sec	19
7	+0.42812794 + 0.40742872i	+1.84601074 + 1.75675939i	0.101sec	28
8	−0.19134172i	−1.00000000i	0.158sec	44
9	+0.16122984	+1.00000000	0.219sec	60
10	−0.36180340 − 0.52573111i	−2.61803399 − 3.80422607i	0.310sec	85
11	−0.01709645 − 0.33745871i	−0.14231484 − 2.80908543i	0.405sec	110
12	−0.78867513 − 0.18301270i	−7.46410162 − 1.73205081i	0.537sec	146
13	+0.37057660 + 0.29130784i	+3.94787634 + 3.10339978i	0.674sec	182
14	−0.68363360 + 0.02926415i	−8.12833414 + 0.34794774i	0.855sec	231
15	+0.39786186 + 0.87662266i	+5.24063637 + 11.54687362i	1.042sec	280
16	+0.04877258 − 0.04877258i	+0.70710678 − 0.70710678i	1.280sec	344
17	+0.06794608 − 0.02632245i	+1.07807180 − 0.41764724i	1.526sec	408
18	+0.05788273	+1.00000000	1.829sec	489
19	+0.12077506 − 0.98833582i	+2.26164019 − 18.50762886i	2.140sec	570
20	−0.65642996 + 0.15590452i	−13.26954255 + 3.15156497i	2.516sec	670
21	+0.24757870 − 0.43688944i	+5.38267899 − 9.49853766i	2.901sec	770
22	−0.64735707 + 0.44552454i	−15.08655417 + 10.38287899i	3.358sec	891
23	+0.14404601 + 0.27366691i	+3.58739662 + 6.81554294i	3.825sec	1012
24	−0.02388795 + 0.60181010i	−0.63397460 + 15.97174719i	4.367sec	1156
25	+0.13820100 − 0.13492161i	+3.89852170 − 3.80601315i	4.927sec	1300

TS-code:

```
dabcgefjhimklonqp
jkfedcpglaqinmhob
lhegcbmfqojpdikan
```

Table 62: Values of the quantum invariants for the 3-manifold induced by $7_2(0)$

r	$S^1 \times S^2 - normalization$	$S^3 - normalization$	$C^{++}, 486$ time	$r-$ states
3	$+1.00000000$	$+1.41421356$	$0.010sec$	2
4			$0.022sec$	6
5	$+0.69098301 + 0.95105652i$	$+1.85874017 + 2.55833637i$	$0.039sec$	10
6	$+1.00000000$	$+3.46410162$	$0.072sec$	19
7	$+0.09903113 + 0.43388374i$	$+0.42700444 + 1.87082869i$	$0.108sec$	28
8	$+1.00000000$	$+5.22625186$	$0.169sec$	44
9	$+0.17364818 - 0.98480775i$	$+1.07702256 - 6.10809849i$	$0.234sec$	60
10	$-0.11803399 - 0.36327126i$	$-0.85410197 - 2.62865556i$	$0.331sec$	85
11	$-0.18639280 - 0.21510876i$	$-1.55157737 - 1.79061574i$	$0.433sec$	110
12	$-0.36602540 - 0.36602540i$	$-3.46410162 - 3.46410162i$	$0.573sec$	146
13	$-1.46796331 - 0.23931566i$	$-15.63870381 - 2.54950976i$	$0.752sec$	182
14	$-0.67844793 + 1.02261879i$	$-8.06667705 + 12.15883360i$	$0.954sec$	231
15	$-0.16913061 + 0.12288058i$	$-2.22778834 + 1.61858298i$	$1.116sec$	280
16	$-0.54119610 + 1.30656296i$	$-7.84628224 + 18.94260101i$	$1.370sec$	344
17	$+0.80483133 + 0.83722036i$	$+12.76991859 + 13.28382166i$	$1.631sec$	408
18	$-0.11334080 + 0.40522291i$	$-1.95811093 + 7.00075722i$	$1.954sec$	489
19	$+0.98873553 + 0.51345701i$	$+18.51511383 + 9.61502311i$	$2.288sec$	570
20	$+0.69098301 - 0.95105652i$	$+13.96802247 - 19.22533359i$	$2.688sec$	670
21	$-0.50140202i$	$-10.90112413i$	$3.100sec$	770
22	$-0.14443723 - 0.36192200i$	$-3.36608673 - 8.43453503i$	$3.587sec$	891
23	$-0.34852731 - 0.54199005i$	$-8.67990531 - 13.49800185i$	$4.086sec$	1012
24	$-1.46592583 - 0.60720636i$	$-38.90495797 - 16.11496124i$	$4.659sec$	1156
25	$-0.96115017 + 0.87859206i$	$-27.11315192 + 24.78426463i$	$5.262sec$	1300
26	$-0.23434678 + 0.14900212i$	$-7.00989383 + 4.45702333i$	$5.948sec$	1469
27	$-0.76764351 + 1.26579219i$	$-24.29521539 + 40.06116569i$	$6.646sec$	1638

TS-code:

$dabcgefjhimklpnorq$
$joledkqgrpicnmbahf$
$mjiplhrfnbdoqcekag$

Table 63: Values of the quantum invariants for the 3-manifold induced by $7_2(1)$

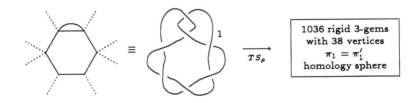

1036 rigid 3-gems
with 38 vertices
$\pi_1 = \pi_1'$
homology sphere

r	$S^1 \times S^2 -$ *normalization*	$S^3 -$ *normalization*	$C^{++}, 486$ *time*	$r-$ *states*
3	+0.70710678	+1.00000000	0.011sec	2
4	−0.50000000	−1.00000000	0.024sec	6
5	−0.04387890 + 0.57206140i	−0.11803399 + 1.53884177i	0.041sec	10
6	+0.28867513	+1.00000000	0.076sec	19
7	−0.40516058 + 0.14540963i	−1.74697960 + 0.62698017i	0.119sec	28
8	+0.19134172i	+1.00000000i	0.186sec	44
9	−0.23846788 − 0.23076559i	−1.47905547 − 1.43128334i	0.248sec	60
10	−0.32532889 + 0.05020285i	−2.35410197 + 0.36327126i	0.351sec	85
11	−0.34819796 − 0.44797814i	−2.89848148 − 3.72907509i	0.458sec	110
12	−0.47168784	−4.46410162	0.607sec	146
13	+0.53280602 + 0.03838045i	+5.67616064 + 0.40887979i	0.762sec	182
14	−0.26105147 − 0.60323838i	−3.10387547 − 7.17244313i	0.967sec	231
15	−0.72113243 + 0.08729741i	−9.49875641 + 1.14988147i	1.177sec	280
16	+0.12343041 − 0.30178463i	+1.78949898 − 4.37528542i	1.446sec	344
17	+0.24635023 − 0.37866392i	+3.90873502 − 6.00810038i	1.723sec	408
18	+0.45694939 − 0.23040125i	+7.89439999 − 3.98048369i	2.065sec	489
19	−0.34519478 − 0.29404101i	−6.46413566 − 5.50622759i	2.416sec	570
20	−0.04662265 + 0.21637176i	−0.94246348 + 4.37389284i	2.840sec	670
21	+0.57917909 − 0.60151706i	+12.59209757 − 13.07775362i	3.274sec	770
22	−0.01153966 − 0.26386128i	−0.26892984 − 6.14924530i	3.789sec	891
23	+0.51126804 + 0.15675394i	+12.73288501 + 3.90388154i	4.314sec	1012
24	+0.16430886 − 0.38658043i	+4.36067709 − 10.25965657i	4.927sec	1156
25	+0.08016654 + 0.21646002i	+2.26142345 + 6.10613587i	5.556sec	1300

TS-code:

dabcgef jhilkomnqpsr
joqedcrgmliaksbnf hp
ophjsnbcf kdmlirgaqe

14.11   Knot 7_3

Table 64: Values of the quantum invariants for the 3-manifold induced by $7_3(7)$

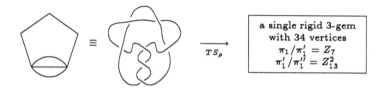

a single rigid 3-gem
with 34 vertices
$\pi_1/\pi_1' = Z_7$
$\pi_1'/\pi_1'' = Z_{13}^2$

r	$S^1 \times S^2$ – normalization	S^3 – normalization	C^{++}, 486 time	$r-$ states
3	$+0.70710678i$	$+1.00000000i$	$0.008sec$	2
4	$-0.35355339 - 0.35355339i$	$-0.70710678 - 0.70710678i$	$0.023sec$	10
5	$+0.48859877 - 0.07099756i$	$+1.31432778 - 0.19098301i$	$0.050sec$	18
6	$+0.28867513$	$+1.00000000$	$0.121sec$	45
7	$-0.38257600 + 0.22737165i$	$-1.64959896 + 0.98038567i$	$0.215sec$	72
8	$+0.32322330 - 0.13388348i$	$+1.68924640 - 0.69970877i$	$0.400sec$	136
9	$+0.10797778 + 0.76090935i$	$+0.66971338 + 4.71940764i$	$0.628sec$	200
10	$-0.23368810 - 0.06937864i$	$-1.69098301 - 0.50202854i$	$1.011sec$	325
11	$-0.01677795 - 0.23847078i$	$-0.13966358 - 1.98508667i$	$1.463sec$	450
12	$-0.11475407 + 0.37357311i$	$-1.08604416 + 3.53553391i$	$2.151sec$	666
13	$-0.44594718 - 0.13493594i$	$-4.75082437 - 1.43751764i$	$2.941sec$	882
14	$+0.33733552 - 0.20048443i$	$+4.01088513 - 2.38373956i$	$4.066sec$	1225
15	$-0.05491713 + 0.01031601i$	$-0.72336846 + 0.13588245i$	$5.330sec$	1568
16	$-0.33253571 - 0.56821650i$	$-4.82111570 - 8.23802508i$	$7.055sec$	2080
17	$+0.47820522 + 0.00072685i$	$+7.58748013 + 0.01153256i$	$8.953sec$	2592
18	$-0.10799265 + 0.27297768i$	$-1.86571467 + 4.71604734i$	$11.446sec$	3321
19	$+0.23928750 - 0.11040508i$	$+4.48091042 - 2.06745128i$	$14.170sec$	4050
20	$+0.34148320 - 0.05958030i$	$+6.90298464 - 1.20439869i$	$17.640sec$	5050
21	$-0.38352294 + 0.50473739i$	$-8.33828148 + 10.97363933i$	$21.396sec$	6050
22	$+0.03111549 - 0.29570429i$	$+0.72514159 - 6.89134164i$	$26.074sec$	7381
23	$+0.04663955 + 0.72545591i$	$+1.16153554 + 18.06713097i$	$31.096sec$	8712
24	$-0.17659830 - 0.19260249i$	$-4.68683302 - 5.11157627i$	$37.232sec$	10440

TS-code:

$$
\begin{array}{l}
dabcgefjhimklonqp \\
jqlodnhgmafcpkeib \\
lfqhnpamcojgdekbi
\end{array}
$$

Table 65: Values of the quantum invariants for the 3-manifold induced by $7_3(-1)$

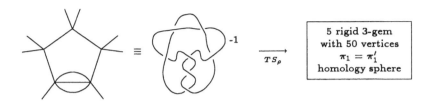

r	$S^1 \times S^2 - normalization$	$S^3 - normalization$	$C^{++}, 486$ time	$r-$ states
3	$+0.70710678$	$+1.00000000$	$0.011 sec$	2
4	-0.50000000	-1.00000000	$0.041 sec$	10
5	$-0.30075048 - 0.48859877i$	$-0.80901699 - 1.31432778i$	$0.082 sec$	18
6	$+0.28867513$	$+1.00000000$	$0.200 sec$	45
7	$-0.17778894 + 0.32673246i$	$-0.76659393 + 1.40881165i$	$0.352 sec$	72
8	$-0.34985438i$	$-1.82842712i$	$0.660 sec$	136
9	$-0.24998803 - 0.01684133i$	$-1.55050720 - 0.10445544i$	$0.976 sec$	200
10	$-0.14827791 - 0.29389263i$	$-1.07294902 - 2.12662702i$	$1.579 sec$	325
11	$-0.07866918 + 0.44548958i$	$-0.65486073 + 3.70835977i$	$2.246 sec$	450
12	$-0.10566243 + 0.36602540i$	$-1.00000000 + 3.46410162i$	$3.315 sec$	666
13	$+0.18615845 - 0.03792027i$	$+1.98320814 - 0.40397726i$	$4.478 sec$	882
14	$+0.41168796 + 0.82977036i$	$+4.89492808 + 9.86588532i$	$6.208 sec$	1225
15	$-0.86066852 - 0.14520341i$	$-11.33672576 - 1.91261926i$	$8.065 sec$	1568
16	$+0.73134925 - 0.22199613i$	$+10.60313008 - 3.21850857i$	$10.689 sec$	2080
17	$-0.39049821 + 0.20748267i$	$-6.19587005 + 3.29203978i$	$13.477 sec$	2592
18	$-0.11383026 - 0.80959806i$	$-1.96656694 - 13.98686821i$	$17.256 sec$	3321
19	$+0.39980459 - 0.11053426i$	$+7.48676184 - 2.06987034i$	$21.244 sec$	4050
20	$-0.05292336 + 0.27887503i$	$-1.06983048 + 5.63737851i$	$26.482 sec$	5050
21	$+0.63610185 + 0.00055604i$	$+13.82967156 + 0.01208900i$	$31.970 sec$	6050
22	$+0.50546576 + 0.08912702i$	$+11.77979955 + 2.07709108i$	$39.001 sec$	7381
23	$+0.61493396 - 0.18055960i$	$+15.31463489 - 4.49675021i$	$46.331 sec$	8712
24	$+0.12822396 - 0.07721350i$	$+3.40300148 - 2.04920877i$	$55.524 sec$	10440
25	$+0.18318012 - 0.02172673i$	$+5.16734065 - 0.61289094i$	$65.066 sec$	12168

TS-code:

$fabcdelghijkomnqpsrutwvyx$
$lkyurgfwptbasdiovqmxnehjc$
$tpvjihuxedcrlsawnmqoykbfg$

Table 66: Values of the quantum invariants for the 3-manifold induced by $7_3(0)$

r	$S^1 \times S^2 - normalization$	$S^3 - normalization$	$C^{++}, 486$ time	$r-$ states
3	$+1.00000000$	$+1.41421356$	$0.010sec$	2
4	$+1.00000000$	$+2.00000000$	$0.037sec$	10
5	$-0.19098301 + 0.58778525i$	$-0.51374315 + 1.58113883i$	$0.073sec$	18
6			$0.178sec$	45
7	$-0.32155207 + 0.15485131i$	$-1.38647471 + 0.66769103i$	$0.301sec$	72
8	-0.41421356	-2.16478440	$0.566sec$	136
9	$+0.38665920 - 0.06817845i$	$+2.39818632 - 0.42286495i$	$0.862sec$	200
10	-0.61803399	-4.47213595	$1.394sec$	325
11	$-0.27532476 - 0.59865916i$	$-2.29186786 - 4.98337929i$	$1.981sec$	450
12	$+1.00000000 + 0.26794919i$	$+9.46410162 + 2.53589838i$	$2.924sec$	666
13	$-0.31128509 + 0.91510366i$	$-3.31622406 + 9.74890516i$	$4.097sec$	882
14	-0.61595706	-7.32366686	$5.680sec$	1225
15	$-0.47007332 - 0.50618662i$	$-6.19180574 - 6.66749019i$	$7.103sec$	1568
16	$+0.54119610 - 0.28269511i$	$+7.84628224 - 4.09852478i$	$9.415sec$	2080
17	$+0.25333636 + 0.55452725i$	$+4.01958098 + 8.79844956i$	$11.862sec$	2592
18	$-1.00000000 + 0.23756470i$	$-17.27631145 + 4.10424172i$	$15.186sec$	3321
19	$+0.88499593 - 0.20740194i$	$+16.57248063 - 3.88381979i$	$18.696sec$	4050
20	$-0.10049177 + 0.48638980i$	$-2.03141225 + 9.83222981i$	$23.302sec$	5050
21	$-1.10487037 - 0.10954489i$	$-24.02130140 - 2.38164669i$	$28.126sec$	6050
22	$+0.53370684 - 0.28618980i$	$+12.43795356 - 6.66960809i$	$34.307sec$	7381
23	$+0.17728882 + 0.07419341i$	$+4.41529293 + 1.84775122i$	$40.754sec$	8712
24	$+0.07542035 + 0.44314565i$	$+2.00161947 + 11.76086991i$	$48.836sec$	10440
25	$-0.35390223 + 0.65776816i$	$-9.98325272 + 18.55502781i$	$57.236sec$	12168

TS-code:
```
fabcdekghijnlmporqtsvuxw
nquwhgfedvpkratlomjbxsic
xkjlpvauqcbdirnsmoehtwfg
```

Table 67: Values of the quantum invariants for the 3-manifold induced by $7_3(1)$

r	$S^1 \times S^2$ − *normalization*	S^3 − *normalization*	C^{++}, 486 *time*	$r-$ *states*
3	$+0.70710678$	$+1.00000000$	$0.010sec$	2
4	-0.50000000	-1.00000000	$0.036sec$	10
5	$+0.11487646 - 0.79056942i$	$+0.30901699 - 2.12662702i$	$0.072sec$	18
6	$+0.28867513$	$+1.00000000$	$0.176sec$	45
7	$-0.03431230 - 0.19019267i$	$-0.14794847 - 0.82007660i$	$0.301sec$	72
8	$+0.57402515i$	$+3.00000000i$	$0.565sec$	136
9	$-0.19777768 + 0.18227290i$	$-1.22668160 + 1.13051587i$	$0.865sec$	200
10	$-0.61180340 + 0.18163563i$	$-4.42705098 + 1.31432778i$	$1.400sec$	325
11	$+0.40818457 + 0.06828559i$	$+3.39782413 + 0.56842531i$	$1.994sec$	450
12	$-0.28867513 + 0.31698730i$	$-2.73205081 + 3.00000000i$	$2.944sec$	666
13	$+0.11562833 - 0.24302237i$	$+1.23182728 - 2.58899853i$	$3.980sec$	882
14	$-0.31029139 + 0.75401104i$	$-3.68933307 + 8.96511473i$	$5.518sec$	1225
15	$+0.12444609 - 0.80896901i$	$+1.63920392 - 10.65573972i$	$7.176sec$	1568
16	$-0.78269703 + 0.69429347i$	$-11.34757220 + 10.06589391i$	$9.511sec$	2080
17	$+0.47537366 - 0.57762797i$	$+7.54255293 - 9.16497890i$	$11.998sec$	2592
18	$-0.55545807 + 0.66551376i$	$-9.59626666 + 11.49762298i$	$15.364sec$	3321
19	$-0.14676138 - 0.05350162i$	$-2.74826136 - 1.00187412i$	$18.924sec$	4050
20	$-0.37999706 + 0.25462076i$	$-7.68153106 + 5.14708543i$	$23.595sec$	5050
21	$+0.20887292 + 0.07046035i$	$+4.54116554 + 1.53189856i$	$28.494sec$	6050
22	$-0.38034805 + 0.07030741i$	$-8.86395117 + 1.63850307i$	$34.766sec$	7381
23	$-0.03684539 + 0.42473124i$	$-0.91761683 + 10.57772761i$	$41.317sec$	8712
24	$-0.26179168 - 0.54217350i$	$-6.94782378 - 14.38902090i$	$49.523sec$	10440
25	$-0.08665304 + 0.21497331i$	$-2.44440172 + 6.06419699i$	$58.057sec$	12168

TS-code:

```
eabcdjfghimklonqpsrutwv
jusofewqrachpmvnltkibdg
qvhgrstcbojpewklnmudfia
```

14.12   Knot 7_4

Table 68: Values of the quantum invariants for the 3-manifold induced by $7_4(-7)$

r	$S^1 \times S^2$ – normalization	S^3 – normalization	$C^{++}, 486$ time	$r-$ states
3	$-0.70710678i$	$-1.00000000i$	$0.007sec$	2
4	$+0.35355339 - 0.35355339i$	$+0.70710678 - 0.70710678i$	$0.012sec$	6
5	$-0.35355339 - 0.85837253i$	$-0.95105652 - 2.30901699i$	$0.021sec$	10
6	$+0.28867513$	$+1.00000000$	$0.040sec$	19
7	$-1.05694620 - 0.52819434i$	$-4.55736202 - 2.27747907i$	$0.061sec$	28
8	$-0.13388348 + 0.32322330i$	$-0.69970877 + 1.68924640i$	$0.095sec$	44
9	$-0.64762169 - 0.00220410i$	$-4.01676068 - 0.01367057i$	$0.133sec$	60
10	$+0.61180340 + 0.60696104i$	$+4.42705098 + 4.39201132i$	$0.188sec$	85
11	$-0.76754664 - 0.26056136i$	$-6.38923829 - 2.16897386i$	$0.246sec$	110
12	$+0.27883877 + 0.27883877i$	$+2.63895843 + 2.63895843i$	$0.327sec$	146
13	$-0.39906917 - 0.08179719i$	$-4.25141720 - 0.87141277i$	$0.412sec$	182
14	$+0.46834140 + 0.34903113i$	$+5.56853167 + 4.14994472i$	$0.522sec$	231
15	$-0.88601527 + 0.07345297i$	$-11.67059315 + 0.96752249i$	$0.637sec$	280
16	$+0.51164258 + 0.62911153i$	$+7.41781409 + 9.12088356i$	$0.783sec$	344
17	$+0.25824121 + 0.28989702i$	$+4.09740411 + 4.59967345i$	$0.936sec$	408
18	$+1.11742584 + 0.22113798i$	$+19.30499677 + 3.82044866i$	$1.122sec$	489
19	$-0.09960496 - 0.46966020i$	$-1.86520766 - 8.79488171i$	$1.315sec$	570
20	$+0.66512530 - 0.34278497i$	$+13.44531635 - 6.92929948i$	$1.547sec$	670
21	$-0.02588357 - 0.64250202i$	$-0.56274205 - 13.96881946i$	$1.785sec$	770
22	$+0.03785002 - 0.35605049i$	$+0.88208882 - 8.29770023i$	$2.067sec$	891
23	$-1.04983872 - 0.41574568i$	$-26.14572901 - 10.35394643i$	$2.357sec$	1012
24	$-0.07349353 + 0.26577677i$	$-1.95048259 + 7.05358610i$	$2.692sec$	1156
25	$-0.22669566 - 0.02668643i$	$-6.39487357 - 0.75279934i$	$3.039sec$	1300

Table 69: Values of the quantum invariants for the 3-manifold induced by $7_4(-1)$

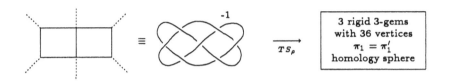

r	$S^1 \times S^2$ − *normalization*	S^3 − *normalization*	$C^{++}, 486$ *time*	$r-$ *states*
3	+0.70710678	+1.00000000	0.010*sec*	2
4	+0.50000000	+1.00000000	0.021*sec*	6
5	+0.37174803 − 0.43701602i	+1.00000000 − 1.17557050i	0.036*sec*	10
6	+0.28867513	+1.00000000	0.067*sec*	19
7	−0.36629931 − 0.51692513i	−1.57941680 − 2.22888825i	0.100*sec*	28
8	−0.34985438	−1.82842712	0.157*sec*	44
9	−0.73628896 + 0.03165136i	−4.56670399 + 0.19631202i	0.217*sec*	60
10	−0.05278640 + 0.32491970i	−0.38196601 + 2.35114101i	0.307*sec*	85
11	+0.49217245 + 0.58309456i	+4.09695894 + 4.85381584i	0.401*sec*	110
12	+0.42264973 − 0.54903811i	+4.00000000 − 5.19615242i	0.531*sec*	146
13	+0.03532824 + 0.08647776i	+0.37636351 + 0.92127651i	0.667*sec*	182
14	−0.23217634 − 0.28470671i	−2.76055314 − 3.38513392i	0.846*sec*	231
15	−0.50768910 − 0.43914970i	−6.68728079 − 5.78447987i	1.032*sec*	280
16	−0.67979997 + 0.73769593i	−9.85576662 + 10.69514441i	1.267*sec*	344
17	+0.50137242 + 0.06182450i	+7.95506425 + 0.98094319i	1.510*sec*	408
18	+0.45501419 − 0.02233812i	+7.86096693 − 0.38592033i	1.810*sec*	489
19	+0.06682106 − 0.15906560i	+1.25129463 − 2.97867080i	2.118*sec*	570
20	+0.21677335 − 0.44979715i	+4.38201078 − 9.09251977i	2.490*sec*	670
21	−0.76122825 + 0.08956107i	−16.55008019 + 1.94717267i	2.870*sec*	770
22	−0.45127763 + 0.10021624i	−10.51695373 + 2.33552352i	3.322*sec*	891
23	+0.07519954 + 0.56175086i	+1.87280841 + 13.99013536i	3.784*sec*	1012
24	+0.31400985 − 0.02477658i	+8.33366869 − 0.65755837i	4.322*sec*	1156
25	+0.53193938 − 0.61677330i	+15.00551489 − 17.39860085i	4.874*sec*	1300

TS-code:
$$dabcgefjhimklpnorq$$
$$mqoedchgnkjirlfbpa$$
$$nrlhpjaokmicqgdefb$$

Table 70: Values of the quantum invariants for the 3-manifold induced by $7_4(0)$

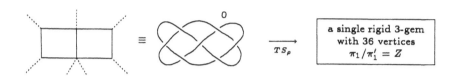

r	$S^1 \times S^2 - normalization$	$S^3 - normalization$	$C^{++}, 486$ time	$r-$ states
3	$+1.00000000$	$+1.41421356$	$0.010sec$	2
4	$+1.00000000$	$+2.00000000$	$0.022sec$	6
5	$+1.19098301 - 0.58778525i$	$+3.20373720 - 1.58113883i$	$0.037sec$	10
6	$+0.50000000 - 0.86602540i$	$+1.73205081 - 3.00000000i$	$0.070sec$	19
7	$-0.70290660 - 1.51597209i$	$-3.03080693 - 6.53659914i$	$0.105sec$	28
8	-1.82842712	-9.55582066	$0.165sec$	44
9	$-0.78698898 + 0.97048085i$	$-4.88116200 + 6.01923836i$	$0.228sec$	60
10	$+0.07294902 + 1.40008449i$	$+0.52786405 + 10.13110656i$	$0.323sec$	85
11	$+0.47481861 + 0.68711368i$	$+3.95250145 + 5.71969541i$	$0.421sec$	110
12	$+0.76794919 + 0.40192379i$	$+7.26794919 + 3.80384758i$	$0.559sec$	146
13	$+0.44320866 - 0.81275230i$	$+4.72164997 - 8.65852187i$	$0.733sec$	182
14	$-0.17187199i$	$-2.04354053i$	$0.930sec$	231
15	$+0.02185240 - 0.28361062i$	$+0.28783980 - 3.73571911i$	$1.088sec$	280
16	$-0.24942916 - 0.54712256i$	$-3.61623375 - 7.93220425i$	$1.333sec$	344
17	$-1.58687707 - 0.38203790i$	$-25.17830782 - 6.06163392i$	$1.590sec$	408
18	$-0.86231110 + 1.11618897i$	$-14.89755516 + 19.28362829i$	$1.906sec$	489
19	$+0.00276524 + 1.34812727i$	$+0.05178203 + 25.24510256i$	$2.231sec$	570
20	$+1.27928811 + 1.10793227i$	$+25.86044082 + 22.39653184i$	$2.622sec$	670
21	$+1.37515472 - 0.24110169i$	$+29.89763042 - 5.24186050i$	$3.023sec$	770
22	$+0.60942383 - 1.98149197i$	$+14.20252592 - 46.17835693i$	$3.499sec$	891
23	$-1.70892447 - 1.32448178i$	$-42.55994308 - 32.98558244i$	$3.985sec$	1012
24	$-1.45421585 + 0.43649585i$	$-38.59418103 + 11.58438741i$	$4.548sec$	1156
25	$-0.72411920 + 1.48010373i$	$-20.42672884 + 41.75234896i$	$5.134sec$	1300

TS-code:

$dabchefgkijnlmporq$
$kpledqnihmarogjbfc$
$mgpiljoanrekhdbqcf$

Table 71: Values of the quantum invariants for the 3-manifold induced by $7_4(1)$

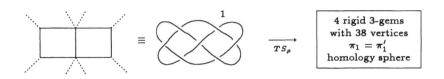

r	$S^1 \times S^2$ – normalization	S^3 – normalization	C^{++}, 486 time	r– states
3	$+0.70710678$	$+1.00000000$	$0.011sec$	2
4	$+0.50000000$	$+1.00000000$	$0.023sec$	6
5	$+0.62861961 - 0.35355339i$	$+1.69098301 - 0.95105652i$	$0.040sec$	10
6	$+0.28867513$	$+1.00000000$	$0.075sec$	19
7	$+0.31924109 - 0.34271535i$	$+1.37651020 - 1.47772697i$	$0.117sec$	28
8	-0.34985438	-1.82842712	$0.183sec$	44
9	$+0.12053964 + 0.51002386i$	$+0.74762613 + 3.16333415i$	$0.244sec$	60
10	$+0.27016261 + 0.44450119i$	$+1.95491503 + 3.21644081i$	$0.345sec$	85
11	$+1.03329022 - 0.10674374i$	$+8.60135016 - 0.88855990i$	$0.450sec$	110
12	$+0.42264973 - 0.54903811i$	$+4.00000000 - 5.19615242i$	$0.596sec$	146
13	$+0.18146563 - 0.02813630i$	$+1.93321397 - 0.29974540i$	$0.749sec$	182
14	$-0.25323052 - 0.23296519i$	$-3.01088513 - 2.76993242i$	$0.950sec$	231
15	$+0.22071882 - 0.10601895i$	$+2.90730835 - 1.39648158i$	$1.156sec$	280
16	$+0.28202482 - 0.29263548i$	$+4.08880695 - 4.24264069i$	$1.420sec$	344
17	$+0.44558331 + 0.30947719i$	$+7.06988209 + 4.91034380i$	$1.692sec$	408
18	$+0.30424575 - 0.49069209i$	$+5.25624426 - 8.47734945i$	$2.028sec$	489
19	$-0.01484288 - 0.64673513i$	$-0.27794860 - 12.11079634i$	$2.372sec$	570
20	$-0.29806981 - 0.74230327i$	$-6.02539528 - 15.00544682i$	$2.788sec$	670
21	$-0.37204297 + 0.30651265i$	$-8.08869222 + 6.66397890i$	$3.213sec$	770
22	$+0.21345773 - 0.01897981i$	$+4.97459852 - 0.44232148i$	$3.720sec$	891
23	$+0.11364841 - 0.16672603i$	$+2.83035916 - 4.15223174i$	$4.235sec$	1012
24	$+0.06096398 - 0.58161763i$	$+1.61795449 - 15.43584881i$	$4.836sec$	1156
25	$-0.54450096 - 0.25789948i$	$-15.35986551 - 7.27510442i$	$5.453sec$	1300

TS-code:
$$dabcgefjhimklonqpsr$$
$$jspmdlhgnacoeqfkrib$$
$$nfsqlraockjegmphdbi$$

14.13 Knot 7_5

Table 72: Values of the quantum invariants for the 3-manifold induced by $7_5(7)$

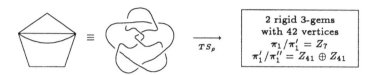

r	$S^1 \times S^2$ − normalization	S^3 − normalization	C^{++}, 486 time	r− states
3	$+0.70710678i$	$+1.00000000i$	$0.007sec$	2
4	$+0.35355339 + 0.35355339i$	$+0.70710678 + 0.70710678i$	$0.022sec$	10
5	$-0.65552404 - 0.07099756i$	$-1.76335576 - 0.19098301i$	$0.051sec$	22
6	$+0.28867513$	$+1.00000000$	$0.125sec$	55
7	$-0.24603621 - 0.05615616i$	$-1.06086391 - 0.24213526i$	$0.241sec$	106
8	$+0.07322330 + 0.17677670i$	$+0.38268343 + 0.92387953i$	$0.460sec$	208
9	$+0.30506070 + 0.23652962i$	$+1.89208583 + 1.46703377i$	$0.768sec$	352
10	$-0.42467111 + 0.09308121i$	$-3.07294902 + 0.67354196i$	$1.283sec$	601
11	$-0.72704477 - 0.27475805i$	$-6.05209122 - 2.28715043i$	$1.958sec$	934
12	$+0.63239216 - 0.37357311i$	$+5.98502365 - 3.53553391i$	$2.980sec$	1450
13	$+0.38477357 - 0.09874608i$	$+4.09912139 - 1.05197503i$	$4.277sec$	2116
14	$-0.02973478 + 0.27069230i$	$-0.35354348 + 3.21850393i$	$6.120sec$	3079
15	$-0.02878267 + 0.02353265i$	$-0.37912530 + 0.30997205i$	$8.375sec$	4278
16	$-0.07137883 + 0.05697164i$	$-1.03485308 + 0.82597703i$	$11.447sec$	5932
17	$-0.11585916 + 0.01123008i$	$-1.83828831 + 0.17818291i$	$15.132sec$	7940
18	$+0.51388522 + 0.02404990i$	$+8.87804107 + 0.41549360i$	$19.966sec$	10601
19	$-0.47237456 + 0.01037714i$	$-8.84571091 + 0.19432288i$	$25.648sec$	13772
20	$-0.58798659 + 0.58426913i$	$-11.88597965 + 11.81083221i$	$32.896sec$	17850
21	$+0.35163011 - 0.91894937i$	$+7.64489045 - 19.97913999i$	$41.311sec$	22624
22	$-0.03519550 - 0.13058760i$	$-0.82022549 - 3.04332340i$	$51.830sec$	28623
23	$+0.58053428 + 0.32739653i$	$+14.45792736 + 8.15365332i$	$63.839sec$	35550
24	$-0.29283605 - 0.13310915i$	$-7.77172617 - 3.53265209i$	$78.555sec$	44076
25	$+0.08939050 + 0.50718478i$	$+2.52162283 + 14.30721062i$	$95.267sec$	53814

TS-code:

> $dabcgefjhimklpnosqrut$
> $muqndphgfkjraecsotilb$
> $rnlsctqfoapebmiugjdhk$

Table 73: Values of the quantum invariants for the 3-manifold induced by $7_5(-1)$

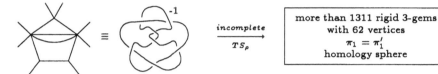

incomplete
$\xrightarrow{TS_\rho}$

more than 1311 rigid 3-gems
with 62 vertices
$\pi_1 = \pi_1'$
homology sphere

r	$S^1 \times S^2$ − normalization	S^3 − normalization	$C^{++}, 486$ time	$r-$ states
3	$+0.70710678$	$+1.00000000$	$0.011sec$	2
4	$+0.50000000$	$+1.00000000$	$0.040sec$	10
5	$-0.30075048 + 0.65552404i$	$-0.80901699 + 1.76335576i$	$0.077sec$	18
6	$+0.28867513$	$+1.00000000$	$0.190sec$	45
7	$+0.33311052 + 0.00886983i$	$+1.43631267 + 0.03824512i$	$0.331sec$	72
8	$+0.19134172$	$+1.00000000$	$0.622sec$	136
9	$-0.16199532 - 0.32190203i$	$-1.00474774 - 1.99654128i$	$0.913sec$	200
10	$-0.33926091 + 0.39429833i$	$-2.45491503 + 2.85316955i$	$1.478sec$	325
11	$+0.24662436 + 0.39421099i$	$+2.05295901 + 3.28150471i$	$2.096sec$	450
12	$+0.33771324 + 0.13397460i$	$+3.19615242 + 1.26794919i$	$3.094sec$	666
13	$-0.03458281 + 0.09182709i$	$-0.36842228 + 0.97826473i$	$4.173sec$	882
14	$-0.39987830 - 0.42646664i$	$-4.75451235 - 5.07064508i$	$5.786sec$	1225
15	$-0.01183986 - 0.31710615i$	$-0.15595466 - 4.17692223i$	$7.505sec$	1568
16	$-0.29895275 - 0.24845948i$	$-4.33422872 - 3.60217527i$	$9.948sec$	2080
17	$+0.59722706 - 0.10728133i$	$+9.47594931 - 1.70218754i$	$12.533sec$	2592
18	$+0.39367712 + 0.45955045i$	$+6.80128850 + 7.93933662i$	$16.046sec$	3321
19	$-0.82024195 + 1.04234938i$	$-15.35989409 + 19.51908968i$	$19.742sec$	4050
20	$-0.04722035 - 0.91206494i$	$-0.95454585 - 18.43713008i$	$24.606sec$	5050
21	$+0.28542745 - 0.38136907i$	$+6.20555950 - 8.29145353i$	$29.693sec$	6050
22	$+0.14799511 + 0.38243511i$	$+3.44900265 + 8.91258971i$	$36.222sec$	7381
23	$-0.33793944 - 0.50532673i$	$-8.41621933 - 12.58491949i$	$43.018sec$	8712
24	$-0.00329505 + 0.04445031i$	$-0.08744913 + 1.17968968i$	$51.542sec$	10440
25	$+0.11185810 + 0.18810686i$	$+3.15541287 + 5.30631964i$	$60.394sec$	12168

Table 74: Values of the quantum invariants for the 3-manifold induced by $7_5(0)$

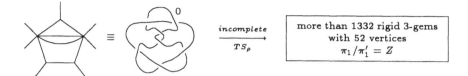

r	$S^1 \times S^2 - normalization$	$S^3 - normalization$	$C^{++}, 486$ time	$r-$ states
3	+1.00000000	+1.41421356	0.011sec	2
4	+1.00000000	+2.00000000	0.038sec	10
5	−0.19098301 + 0.58778525i	−0.51374315 + 1.58113883i	0.077sec	18
6			0.189sec	45
7	−0.32155207 + 0.15485131i	−1.38647471 + 0.66769103i	0.325sec	72
8	−0.41421356	−2.16478440	0.609sec	136
9	+0.38665920 − 0.06817845i	+2.39818632 − 0.42286495i	0.937sec	200
10	−0.61803399	−4.47213595	1.514sec	325
11	−0.27532476 − 0.59865916i	−2.29186786 − 4.98337929i	2.165sec	450
12	+1.00000000 + 0.26794919i	+9.46410162 + 2.53589838i	3.194sec	666
13	−0.31128509 + 0.91510366i	−3.31622406 + 9.74890516i	4.482sec	882
14	−0.61595706	−7.32366686	6.209sec	1225
15	−0.47007332 − 0.50618662i	−6.19180574 − 6.66749019i	7.819sec	1568
16	+0.54119610 − 0.28269511i	+7.84628224 − 4.09852478i	10.357sec	2080
17	+0.25333636 + 0.55452725i	+4.01958098 + 8.79844956i	13.086sec	2592
18	−1.00000000 + 0.23756470i	−17.27631145 + 4.10424172i	16.745sec	3321
19	+0.88499593 − 0.20740194i	+16.57248063 − 3.88381979i	20.660sec	4050
20	−0.10049177 + 0.48638980i	−2.03141225 + 9.83229981i	25.739sec	5050
21	−1.10487037 − 0.10954489i	−24.02130140 − 2.38164669i	31.124sec	6050
22	+0.53370684 − 0.28618980i	+12.43795356 − 6.66960809i	37.950sec	7381
23	+0.17728882 + 0.07419341i	+4.41529293 + 1.84775122i	45.150sec	8712
24	+0.07542035 + 0.44314565i	+2.00161947 + 11.76086991i	54.085sec	10440
25	−0.35390223 + 0.65776816i	−9.98325272 + 18.55502781i	63.465sec	12168
26	−0.32003315 − 0.78751762i	−9.57298575 − 23.55660651i	74.911sec	14365
27	+0.42964477 − 0.09460793i	+13.59786425 − 2.99425459i	86.874sec	16562
28	−0.29162241 + 0.87938509i	−9.74551199 + 29.38751507i	101.261sec	19306
29	−0.22185427 − 0.34866564i	−7.81357353 − 12.27979345i	116.243sec	22050
30	+0.41801156 + 0.02569865i	+15.48814336 + 0.95218512i	134.022sec	25425

Table 75: Values of the quantum invariants for the 3-manifold induced by $7_5(1)$

| incomplete $\xrightarrow{TS_\rho}$ | more than 1337 rigid 3-gems with 50 vertices $\pi_1 = \pi_1'$ homology sphere |

r	$S^1 \times S^2$ – normalization	S^3 – normalization	$C^{++}, 486$ time	$r-$ states
3	+0.70710678	+1.00000000	0.008sec	2
4	−0.50000000	−1.00000000	0.028sec	10
5	−0.55762205 − 0.13504538i	−1.50000000 − 0.36327126i	0.059sec	18
6	+0.28867513	+1.00000000	0.143sec	45
7	+0.24579004 + 0.53509868i	+1.05980247 + 2.30724933i	0.252sec	72
8	+0.03282905i	+0.17157288i	0.471sec	136
9	+0.40307460 − 0.28058412i	+2.50000000 − 1.74027413i	0.736sec	200
10	+0.29270510 − 0.15060856i	+2.11803399 − 1.08981379i	1.188sec	325
11	−0.13705428 + 0.50030468i	−1.14087201 + 4.16465354i	1.715sec	450
12	−0.47168784 − 0.26794919i	−4.46410162 − 2.53589838i	2.526sec	666
13	−0.33049884 + 0.19440845i	−3.52091463 + 2.07109820i	3.449sec	882
14	−0.05772540 − 0.11447707i	−0.68634916 − 1.36112077i	4.774sec	1225
15	−0.08602283 − 0.04873387i	−1.13309271 − 0.64192255i	6.250sec	1568
16	+0.01609820 − 0.26250447i	+0.23339228 − 3.80580012i	8.274sec	2080
17	+0.04180166 − 0.33448740i	+0.66324925 − 5.30717023i	10.499sec	2592
18	−0.20986350 − 0.49582744i	−3.62566720 − 8.56606928i	13.429sec	3321
19	−0.21750013 − 0.48042269i	−4.07291890 − 8.99642068i	16.616sec	4050
20	−0.07522085 + 0.07862587i	−1.52056775 + 1.58939930i	20.693sec	5050
21	−0.55709160 − 0.13284375i	−12.11188703 − 2.88819376i	25.088sec	6050
22	+0.05743182 + 0.91973380i	+1.33843951 + 21.43425080i	30.586sec	7381
23	−0.90818873 + 0.31360483i	−22.61800418 + 7.81017778i	36.462sec	8712
24	+0.16158241 − 0.10027139i	+4.28831857 − 2.66115384i	43.665sec	10440
25	+0.28653428 + 0.66896470i	+8.08286550 + 18.87087164i	51.336sec	12168

14.14 Knot 7_6

Table 76: Values of the quantum invariants for the 3-manifold induced by $7_6(3)$

$$\boxed{\begin{array}{c} \text{2 rigid 3-gems} \\ \text{with 46 vertices} \\ \pi_1/\pi_1' = Z_3 \\ \pi_1'/\pi_1'' = Z_{11}^2 \end{array}}$$

r	$S^1 \times S^2 - normalization$	$S^3 - normalization$	$C^{++}, 486$ time	$r-$ states
3	$+0.70710678i$	$+1.00000000i$	$0.007sec$	2
4	$+0.35355339 - 0.35355339i$	$+0.70710678 - 0.70710678i$	$0.023sec$	10
5	$+0.27009076 + 0.22975292i$	$+0.72654253 + 0.61803399i$	$0.058sec$	26
6	$-0.50000000i$	$-1.73205081i$	$0.141sec$	65
7	$+0.48812497 + 0.19418285i$	$+2.10470714 + 0.83728154i$	$0.305sec$	152
8	$+0.11611652 - 0.28033009i$	$+0.60685420 - 1.46507563i$	$0.605sec$	312
9	$+0.61215271 + 0.38168965i$	$+3.79677054 + 2.36736354i$	$1.110sec$	606
10	$-0.02016261 + 0.12410828i$	$-0.14589803 + 0.89805595i$	$1.949sec$	1101
11	$-0.20769661 - 0.38371354i$	$-1.72891532 - 3.19412147i$	$3.243sec$	1906
12	$+0.42290374 + 0.51763809i$	$+4.00240401 + 4.89897949i$	$5.206sec$	3144
13	$+0.06085106 - 0.10796752i$	$+0.64826664 - 1.15021409i$	$8.094sec$	5020
14	$-0.00908804 + 0.02871438i$	$-0.10805583 + 0.34141107i$	$12.225sec$	7727
15	$-0.23556947 - 0.21270506i$	$-3.10292107 - 2.80175106i$	$17.972sec$	11588
16	$+0.38456918 - 0.01556260i$	$+5.57549901 - 0.22562725i$	$25.894sec$	16930
17	$-0.01491000 - 0.29433185i$	$-0.23657070 - 4.67003912i$	$36.554sec$	24212
18	$+0.80473520 + 0.26109181i$	$+13.90285599 + 4.51070343i$	$50.718sec$	33915
19	$-0.32822984 + 0.06185083i$	$-6.14644934 + 1.15822201i$	$69.189sec$	46724
20	$+0.37824830 - 0.38253091i$	$+7.64618050 - 7.73275228i$	$93.026sec$	63292
21	$+0.16716136 + 0.61447353i$	$+3.63430265 + 13.35944405i$	$123.506sec$	84552
22	$+0.84628519 + 0.03897574i$	$+19.72254241 + 0.90832352i$	$162.053sec$	111437
23	$-0.48639181 + 0.41979116i$	$-12.11335430 + 10.45469728i$	$210.069sec$	145154
24	$+0.33976829 + 0.26396195i$	$+9.01728507 + 7.005542161i$	$269.590sec$	186934

TS-code:
$$\boxed{\begin{array}{l} dabchefgkijnlmqopsrutwv \\ kpswdrqihlnjgufbtovmace \\ qvtspmkulcgrfonewidjbha \end{array}}$$

Table 77: Values of the quantum invariants for the 3-manifold induced by $7_6(-1)$

r	$S^1 \times S^2 -$ *normalization*	$S^3 -$ *normalization*	$C^{++}, 486$ *time*	$r-$ *states*
3	$+0.70710678$	$+1.00000000$	$0.009sec$	2
4	-0.50000000	-1.00000000	$0.032sec$	10
5	$-0.45950584 + 0.43701602i$	$-1.23606798 + 1.17557050i$	$0.066sec$	18
6	$+0.28867513$	$+1.00000000$	$0.162sec$	45
7	$+0.51544842 - 0.45932481i$	$+2.22252093 - 1.98052603i$	$0.282sec$	72
8	$-0.34985438i$	$-1.82842712i$	$0.528sec$	136
9	$-0.30080888 + 0.11515256i$	$-1.86571467 + 0.71421370i$	$0.819sec$	200
10	$+0.02016261 + 0.40162283i$	$+0.14589803 + 2.90617011i$	$1.324sec$	325
11	$-0.38937105 + 0.17921910i$	$-3.24121594 + 1.49186179i$	$1.901sec$	450
12	$-0.10566243 + 0.09807621i$	$-1.00000000 + 0.92820323i$	$2.803sec$	666
13	$-0.33174983 + 0.52680079i$	$-3.53424182 + 5.61218490i$	$3.814sec$	882
14	$+0.04375801 - 0.28272542i$	$+0.52027831 - 3.36157657i$	$5.283sec$	1225
15	$+0.09792624 - 0.52888756i$	$+1.28988442 - 6.96650691i$	$6.901sec$	1568
16	$-0.17941022 - 0.10640330i$	$-2.60109644 - 1.54263921i$	$9.140sec$	2080
17	$-0.11581077 + 0.62768000i$	$-1.83752053 + 9.95913323i$	$11.574sec$	2592
18	$+0.45070513 + 0.68878760i$	$+7.78652215 + 11.89970915i$	$14.810sec$	3321
19	$+0.44358689 - 0.05506707i$	$+8.30663161 - 1.03118883i$	$18.297sec$	4050
20	$-0.26788070 - 0.37413881i$	$-5.41513113 - 7.56310821i$	$22.794sec$	5050
21	$-0.33648689 - 0.76904735i$	$-7.31565734 - 16.72007741i$	$27.600sec$	6050
22	$+1.00462518 + 0.27089986i$	$+23.41263098 + 6.31327827i$	$33.655sec$	7381
23	$+0.60290642 + 0.71897928i$	$+15.01509482 + 17.90583366i$	$40.081sec$	8712
24	$-0.10522928 + 0.23708539i$	$-2.79273386 + 6.29213083i$	$48.017sec$	10440
25	$-0.47166249 - 0.21470041i$	$-13.30516010 - 6.05649890i$	$56.387sec$	12168

TS-code:
$$dabchefgkijnlmporqtsvuxw$$
$$htledjsaprckuqnioxgvwbmf$$
$$jwfskpbnhmqxaiucvdrotegl$$

Table 78: Values of the quantum invariants for the 3-manifold induced by $7_6(0)$

r	$S^1 \times S^2$ – normalization	S^3 – normalization	$C^{++}, 486$ time	$r-$ states
3	+1.00000000	+1.41421356	0.009sec	2
4			0.031sec	10
5	−0.30901699 + 0.22451399i	−0.83125388 + 0.60394129i	0.066sec	18
6			0.160sec	45
7	+0.48038567 − 0.15485131i	+2.07133666 − 0.66769103i	0.281sec	72
8	+0.58578644 − 0.41421356i	+3.06146746 − 2.16478440i	0.525sec	136
9	+0.47430773 − 0.56525794i	+2.94181103 − 3.50591386i	0.820sec	200
10	+0.23606798	+1.70820393	1.323sec	325
11	+0.05054732 + 0.23552765i	+0.42076776 + 1.96058741i	1.909sec	450
12	−0.46410162 + 0.73205081i	−4.39230485 + 6.92820323i	2.813sec	666
13	−0.30940895 − 0.36971826i	−3.29623699 − 3.93873217i	3.838sec	882
14	+0.23005368 − 0.07415732i	+2.73531489 − 0.88172302i	5.315sec	1225
15	+0.03449508 + 0.08374867i	+0.45436916 + 1.10313749i	6.955sec	1568
16	−0.25925355 + 0.33257392i	−3.75866806 + 4.82166969i	9.210sec	2080
17	−0.49534888 + 0.41203758i	−7.85949133 + 6.53762618i	11.676sec	2592
18	−0.47384090 − 0.16003503i	−8.18622297 − 2.76481496i	14.937sec	3321
19	−0.03832861 − 0.58458675i	−0.71774359 − 10.94700248i	18.476sec	4050
20	+0.14200573 − 0.33830457i	+2.87060502 − 6.83872949i	23.014sec	5050
21	−0.92456286 − 0.66134456i	−20.10118440 − 14.37848030i	27.887sec	6050
22	−0.07165301 − 0.77545059i	−1.66986214 − 18.07175339i	34.000sec	7381
23	−0.30982208 + 0.25208189i	−7.71597016 + 6.27797840i	40.515sec	8712
24	−0.34822907 + 0.22149255i	−9.24182994 + 5.87830437i	48.531sec	10440
25	+0.02777504 − 0.44212530i	+0.78350802 − 12.47194330i	57.025sec	12168
26	+0.70321109 − 0.62055913i	+21.03478913 − 18.56246368i	67.315sec	14365
27	+0.66620468 − 0.40650536i	+21.08476926 − 12.86552309i	78.142sec	16562
28	+0.14723665 − 0.26215975i	+4.92039175 − 8.76092119i	91.098sec	19306
29	−0.18153581 + 0.10861454i	−6.39358172 + 3.82533868i	104.676sec	22050
30	−0.18879475 − 0.05772765i	−6.99521366 − 2.13892200i	120.691sec	25425

Table 79: Values of the quantum invariants for the 3-manifold induced by $7_6(1)$

	a single rigid 3-gems
	with 46 vertices
	$\pi_1 = \pi_1'$
	homology sphere

r	$S^1 \times S^2$ − *normalization*	S^3 − *normalization*	$C^{++}, 486$ *time*	$r−$ *states*
3	+0.70710678	+1.00000000	0.008*sec*	2
4	−0.50000000	−1.00000000	0.028*sec*	10
5	−0.55762205 − 0.13504538i	−1.50000000 − 0.36327126i	0.059*sec*	18
6	+0.28867513	+1.00000000	0.143*sec*	45
7	+0.24579004 + 0.53509868i	+1.05980247 + 2.30724933i	0.252*sec*	72
8	+0.03282905i	+0.17157288i	0.471*sec*	136
9	+0.40307460 − 0.28058412i	+2.50000000 − 1.74027413i	0.736*sec*	200
10	+0.29270510 − 0.15060856i	+2.11803399 − 1.08981379i	1.188*sec*	325
11	−0.13705428 + 0.50030468i	−1.14087201 + 4.16465354i	1.715*sec*	450
12	−0.47168784 − 0.26794919i	−4.46410162 − 2.53589838i	2.526*sec*	666
13	−0.33049884 + 0.19440845i	−3.52091463 + 2.07109820i	3.449*sec*	882
14	−0.05772540 − 0.11447707i	−0.68634916 − 1.36112077i	4.774*sec*	1225
15	−0.08602283 − 0.04873387i	−1.13309271 − 0.64192255i	6.250*sec*	1568
16	+0.01609820 − 0.26250447i	+0.23339228 − 3.80580012i	8.274*sec*	2080
17	+0.04180166 − 0.33448740i	+0.66324925 − 5.30717023i	10.499*sec*	2592
18	−0.20986350 − 0.49582744i	−3.62566720 − 8.56606928i	13.429*sec*	3321
19	−0.21750013 − 0.48042269i	−4.07291890 − 8.99642068i	16.616*sec*	4050
20	−0.07522085 + 0.07862587i	−1.52056775 + 1.58939930i	20.693*sec*	5050
21	−0.55709160 − 0.13284375i	−12.11188703 − 2.88819376i	25.088*sec*	6050
22	+0.05743182 + 0.91973380i	+1.33843951 + 21.43425080i	30.586*sec*	7381
23	−0.90818873 + 0.31360483i	−22.61800418 + 7.81017778i	36.462*sec*	8712
24	+0.16158241 − 0.10027139i	+4.28831857 − 2.66115384i	43.665*sec*	10440
25	+0.28653428 + 0.66896470i	+8.08286550 + 18.87087164i	51.336*sec*	12168

TS-code:	$eabcdifghljkomnqpsrutwv$
	$loqvfedtjimakrbnuphswgc$
	$mrhptolcbvdqsewkgianjuf$

14.15   Knot 7_7

Table 80: Values of the quantum invariants for the 3-manifold induced by $7_7(-1)$

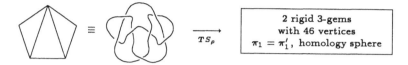

			2 rigid 3-gems with 46 vertices $\pi_1 = \pi_1'$, homology sphere

r	$S^1 \times S^2$ – *normalization*	S^3 – *normalization*	$C^{++}, 486$ *time*	$r-$ *states*
3	$+0.70710678$	$+1.00000000$	$0.007sec$	2
4	-0.50000000	-1.00000000	$0.023sec$	10
5	$-0.45950584 - 0.43701602i$	$-1.23606798 - 1.17557050i$	$0.048sec$	18
6	$+0.28867513$	$+1.00000000$	$0.117sec$	45
7	$+0.83783749 + 0.15252268i$	$+3.61260467 + 0.65765037i$	$0.205sec$	72
8	$-0.19134172i$	$-1.00000000i$	$0.383sec$	136
9	$-0.65490856 - 0.54885000i$	$-4.06195625 - 3.40414649i$	$0.599sec$	200
10	$+0.02016261 - 0.40162283i$	$+0.14589803 - 2.90617011i$	$0.966sec$	325
11	$+0.28654649 - 0.57543623i$	$+2.38528020 - 4.79006616i$	$1.395sec$	450
12	$-0.23963703 + 0.50000000i$	$-2.26794919 + 4.73205081i$	$2.051sec$	666
13	$-0.11977688 + 0.61827994i$	$-1.27602319 + 6.58674284i$	$2.804sec$	882
14	$-0.13262455 + 0.51747569i$	$-1.57689239 + 6.15273343i$	$3.880sec$	1225
15	$+0.02395186 - 0.20607442i$	$+0.31549385 - 2.71441227i$	$5.080sec$	1568
16	$-0.69526639 - 0.01815048i$	$-10.07999939 - 0.26314636i$	$6.724sec$	2080
17	$-0.08127233 + 0.63247280i$	$-1.28951374 + 10.03517862i$	$8.533sec$	2592
18	$+0.83052904 + 0.45356734i$	$+14.34847840 + 7.83597055i$	$10.910sec$	3321
19	$+0.48186993 - 0.24098291i$	$+9.02352177 - 4.51265870i$	$13.504sec$	4050
20	$-0.26329870 - 0.13698532i$	$-5.32250740 - 2.76911882i$	$16.815sec$	5050
21	$-0.14894913 - 0.56975632i$	$-3.23834536 - 12.38723430i$	$20.386sec$	6050
22	$+0.47411934 - 0.35153168i$	$+11.04927630 - 8.19239024i$	$24.854sec$	7381
23	$+0.82669210 - 0.40463693i$	$+20.58836960 - 10.07728837i$	$29.632sec$	8712
24	$-0.19950037 + 0.51610163i$	$-5.29464279 + 13.69708616i$	$35.473sec$	10440
25	$-0.26437537 + 0.28117134i$	$-7.45778321 + 7.93158181i$	$41.720sec$	12168

TS-code:	$dabcgefjhimklpnosqrutwv$ $jwmndphgrvbiuetsolfqcak$ $rgolqtipbujsvmwheadfknc$

Table 81: Values of the quantum invariants for the 3-manifold induced by $7_7(0)$

r	$S^1 \times S^2 - normalization$	$S^3 - normalization$	$C^{++}, 486$ time	$r-$ states
3	+1.00000000	+1.41421356	0.008sec	2
4			0.027sec	10
5	−0.61803399	−1.66250775	0.058sec	18
6			0.141sec	45
7	+1.00000000	+4.31182025	0.253sec	72
8	+1.00000000	+5.22625186	0.471sec	136
9			0.742sec	200
10	−0.61803399	−4.47213595	1.195sec	325
11			1.734sec	450
12	+1.00000000	+9.46410162	2.550sec	666
13	+1.00000000	+10.65333422	3.495sec	882
14			4.832sec	1225
15	−0.61803399	−8.14074369	6.343sec	1568
16			8.391sec	2080
17	+1.00000000	+15.86657740	10.666sec	2592
18	+1.00000000	+17.27631145	13.629sec	3321
19			16.895sec	4050
20	−0.61803399	−12.49337910	21.029sec	5050
21			25.524sec	6050
22	+1.00000000	+23.30484179	31.100sec	7381
23	+1.00000000	+24.90451968	37.116sec	8712
24			44.425sec	10440
25	−0.61803399	−17.43416379	52.278sec	12168

TS-code:
$cabfdeighljknm$
$imlckgfenbhdja$
$kenmbjlcfiagdh$

Table 82: Values of the quantum invariants for the 3-manifold induced by $7_7(1)$

$$R_{28}(172) \in C_{28}(3)$$
6 rigid 3-gems
$$\pi_1 = \pi_1' = \langle 7, 3, 2 \rangle$$
homology sphere

r	$S^1 \times S^2 - normalization$	$S^3 - normalization$	$C^{++}, 486$ time	$r-$ states
3	$+0.70710678$	$+1.00000000$	$0.009sec$	2
4	-0.50000000	-1.00000000	$0.029sec$	10
5	$-0.71637742 - 0.35355339i$	$-1.92705098 - 0.95105652i$	$0.061sec$	18
6	$+0.28867513$	$+1.00000000$	$0.149sec$	45
7	$+0.58547400 + 0.73416118i$	$+2.52445867 + 3.16557105i$	$0.264sec$	72
8	$+0.19134172i$	$+1.00000000i$	$0.494sec$	136
9	$+0.09888884 - 0.87838728i$	$+0.61334080 - 5.44804403i$	$0.774sec$	200
10	$+0.38819660 - 0.50655533i$	$+2.80901699 - 3.66546879i$	$1.249sec$	325
11	$-0.36522184 + 0.40677809i$	$-3.04019223 + 3.38611619i$	$1.806sec$	450
12	$-0.78867513 + 0.18301270i$	$-7.46410162 + 1.73205081i$	$2.660sec$	666
13	$+0.24724342 - 0.10586787i$	$+2.63396677 - 1.12784583i$	$3.635sec$	882
14	$+0.85966457 + 0.41399264i$	$+10.22132448 + 4.92233044i$	$5.033sec$	1225
15	$+0.08219527 + 0.20412415i$	$+1.08267606 + 2.68872324i$	$6.594sec$	1568
16	$-0.26138010 - 0.81695034i$	$-3.78949898 - 11.84417797i$	$8.730sec$	2080
17	$+0.16713092 - 0.59182253i$	$+2.65179560 - 9.39019802i$	$11.079sec$	2592
18	$-0.19211727 + 0.53326852i$	$-3.31907786 + 9.21291311i$	$14.175sec$	3321
19	$-0.67439974 + 0.46666858i$	$-12.62884527 + 8.73886052i$	$17.541sec$	4050
20	$+0.19737596 - 0.16561496i$	$+3.98989828 - 3.34785876i$	$21.852sec$	5050
21	$+0.94865073 + 0.12384016i$	$+20.62488555 + 2.69244420i$	$26.490sec$	6050
22	$+0.14411067 + 0.18165006i$	$+3.35847636 + 4.23332596i$	$32.304sec$	7381
23	$-0.50508391 - 0.67914673i$	$-12.57887210 - 16.91382302i$	$38.508sec$	8712
24	$-0.02388795 - 0.60181010i$	$-0.63397460 - 15.97174719i$	$46.125sec$	10440
25	$-0.01964222 + 0.57823771i$	$-0.55408868 + 16.31154775i$	$54.225sec$	12168

TS-code:
dabcgefjhilknm
jlmedchgnaibfk
nfikmjaecbdhlg

14.16   Links with 2 Components

Table 83: Values of the quantum invariants for the 3-manifold induced by $4_1^2(0,0)$

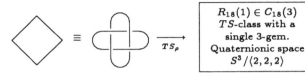

		$R_{18}(1) \in C_{18}(3)$
		TS-class with a
	$\equiv$	single 3-gem.
	$\xrightarrow{TS_\rho}$	Quaternionic space
		$S^3/\langle 2,2,2 \rangle$

$$TS_\rho \uparrow$$

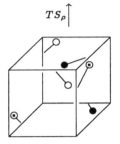

r	$S^1 \times S^2 -$ normalization	$S^3 -$ normalization	$C^{++}, 486$ time	$r-$ states
3	+1.41421356	+2.00000000	0.007sec	4
4	+1.50000000 + 0.50000000i	+3.00000000 + 1.00000000i	0.012sec	10
5	+1.11524410 + 1.14412281i	+3.00000000 + 3.07768354i	0.025sec	20
6	+0.28867513 + 1.50000000i	+1.00000000 + 5.19615242i	0.044sec	35
7	−0.66117206 + 1.26706919i	−2.85085508 + 5.46337461i	0.070sec	56
8	−1.27373392 + 0.46193977i	−6.65685425 + 2.41421356i	0.105sec	84
9	−1.21205194 − 0.55851656i	−7.51754097 − 3.46410162i	0.151sec	120
10	−0.47983739 − 1.27597621i	−3.47213595 − 9.23305061i	0.208sec	165
11	+0.54864472 − 1.30064086i	+4.56704736 − 10.82683953i	0.294sec	220
12	+1.31698730 − 0.60566243i	+12.46410162 − 5.73205081i	0.381sec	286
13	+1.39389429 + 0.43709222i	+14.84962173 + 4.65648949i	0.463sec	364
14	+0.72362901 + 1.25143152i	+8.60387547 + 14.87939376i	0.579sec	455
15	−0.33004844 + 1.37759673i	−4.34739800 + 18.14570407i	0.715sec	560
16	−1.18011531 + 0.73694061i	−17.10935797 + 10.68419374i	0.866sec	680
17	−1.34643031 − 0.31525559i	−21.36324070 − 5.00202729i	1.042sec	816
18	−0.73197896 − 1.18531940i	−12.64589651 − 20.47794707i	1.238sec	969

This manifold has fundamental group the finite quaternions, $Q_8 = \langle 2,2,2 \rangle = \langle a, b, c \mid a^2 = b^2 = c^2 = abc \rangle$. It is homeomorphic to the quotient $S^3/\langle 2,2,2 \rangle$. It is also obtained from a solid cube by identifying the three pairs of opposite faces under a $\pi/2$-rotation, as indicated in the right diagram above. See [LD89], [LD91], [LS92]

$$TS\text{-code:} \quad \begin{array}{|l|} \hline cabfdeigh \\ ihdcbgfea \\ gfhaibdce \\ \hline \end{array}$$

Table 84: Values of the quantum invariants for the 3-manifold induced by $5_1^2(1,0)$
(the simplest projection of Whitehead's link)

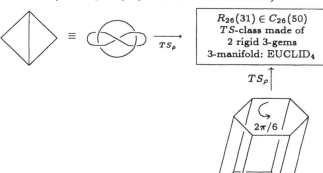

$$R_{26}(31) \in C_{26}(50)$$
TS-class made of
2 rigid 3-gems
3-manifold: EUCLID$_4$

$TS_\rho\uparrow$

$2\pi/6$

r	$S^1 \times S^2$ – *normalization*	S^3 – *normalization*	C^{++}, 486 time	$r-$ states
3	$+1.00000000$	$+1.41421356$	$0.017sec$	4
4			$0.035sec$	12
5	$-0.30901699 - 0.95105652i$	$-0.83125388 - 2.55833637i$	$0.084sec$	28
6	$+0.50000000 - 0.86602540i$	$+1.73205081 - 3.00000000i$	$0.178sec$	57
7			$0.335sec$	104
8	$-0.70710678 - 0.70710678i$	$-3.69551813 - 3.69551813i$	$0.579sec$	176
9	$+0.17364818 - 0.98480775i$	$+1.07702256 - 6.10809849i$	$0.938sec$	280
10			$1.449sec$	425
11	$-0.84125353 - 0.54064082i$	$-7.00279171 - 4.50042096i$	$2.187sec$	620
12	$-1.00000000i$	$-9.46410162i$	$3.118sec$	876
13			$4.263sec$	1204
14	$-0.90096887 - 0.43388374i$	$-10.71242836 - 5.15883360i$	$5.771sec$	1617
15	$-0.10452846 - 0.99452190i$	$-1.37684892 - 13.09984239i$	$7.686sec$	2128
16			$10.011sec$	2752

This manifold can be obtained from a solid hexagonal prism by identifying via translations the three pairs of opposite lateral faces and the bases by a $\pi/3$-rotation in the upper one followed by a translation. It is the euclidean 3-manifold whose fundamental group (see EUCLID$_4$ in [LS92]) is

$$\langle 6,3,2 \rangle = \langle a,b,c \mid a^6 = b^3 = c^2 = abc \rangle.$$

Another blink inducing the same manifold:

TS-code:
| $dabcgefihkjml$ |
| $ijmedckgalhbf$ |
| $lfhkjbaceimgd$ |

Table 84: Values of the quantum invariants for the 3-manifold induced by $6_1^2(0,0)$

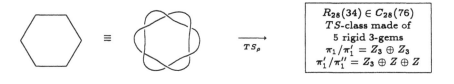

$$R_{28}(34) \in C_{28}(76)$$
TS-class made of
5 rigid 3-gems
$$\pi_1/\pi_1' = Z_3 \oplus Z_3$$
$$\pi_1'/\pi_1'' = Z_3 \oplus Z \oplus Z$$

r	$S^1 \times S^2$ − *normalization*	S^3 − *normalization*	C^{++},486 time	$r-$ states
3	+0.70710678	+1.00000000	0.018sec	4
4	+0.50000000	+1.00000000	0.030sec	10
5	+0.55762205 − 0.57206140i	+1.50000000 − 1.53884177i	0.057sec	20
6	+0.86602540	+3.00000000	0.099sec	35
7	+0.55683417 + 0.40742872i	+2.40096887 + 1.75675939i	0.159sec	56
8	+0.19134172	+1.00000000	0.239sec	84
9	+0.42134852 + 0.35355339i	+2.61334080 + 2.19285330i	0.339sec	120
10	−0.01631190 + 0.63798811i	−0.11803399 + 4.61652531i	0.467sec	165
11	−0.52065882 + 0.02584127i	−4.33408616 + 0.21510876i	0.621sec	220
12	−0.18301270	−1.73205080	0.806sec	286
13	−0.38625062 + 0.08109199i	−4.11485697 + 0.86390004i	1.027sec	364
14	−0.62030082 − 0.60224774i	−7.37531375 − 7.16066445i	1.282sec	455
15	+0.05348379 − 0.64035921i	+0.70448853 − 8.43481145i	1.580sec	560
16	+0.13794969 − 0.34675996i	+2.00000000 − 5.02733949i	1.916sec	680
17	+0.03350291 − 0.69778730i	+0.53157647 − 11.07149619i	2.303sec	816
18	+0.68969262 − 0.43301270i	+11.91534452 − 7.48086230i	2.725sec	969
19	+0.60248079 + 0.16134250i	+11.28208709 + 3.02130827i	3.222sec	1140
20	+0.18600678 − 0.00510508i	+3.76007351 − 0.10319775i	3.749sec	1330
21	+0.49398634 + 0.22933025i	+10.73989758 + 4.98593432i	4.349sec	1540
22	+0.11230802 + 0.65701891i	+2.61732055 + 15.31172177i	4.996sec	1771
23	−0.48884409 + 0.20316958i	−12.17442729 + 5.05984089i	5.699sec	2024

The fundamental group of this manifold is (see [LS92])

$$\langle 3,3,3 \rangle = \langle a,b,c \mid a^3 = b^3 = c^3 \rangle.$$

TS-code:

dabcgefjhilknm
gkindhlfmbjace
ngkimjbedlcfha

Table 85: Values of the quantum invariants for the 3-manifold induced by $6_2^2(0,0)$

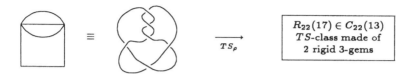

$$R_{22}(17) \in C_{22}(13)$$
TS-class made of
2 rigid 3-gems

r	$S^1 \times S^2 - normalization$	$S^3 - normalization$	$C^{++}, 486$ time	$r-$ states
3	+0.70710678	+1.00000000	0.021 sec	4
4	+0.50000000	+1.00000000	0.070 sec	16
5	+0.97324899	+2.61803399	0.221 sec	44
6	+0.86602540	+3.00000000	0.593 sec	107
7	+0.75304150	+3.24697960	1.355 sec	224
8	+1.11522125	+5.82842712	2.778 sec	432
9	+1.03372096	+6.41147413	5.195 sec	768
10	+0.94721360	+6.85410197	9.097 sec	1293
11	+1.25231353	+10.42455157	15.026 sec	2068
12	+1.18301270	+11.19615242	23.759 sec	3184
13	+1.11008143	+11.82606850	36.169 sec	4732
14	+1.37879492	+16.39373162	53.311 sec	6839
15	+1.31720000	+17.35015835	76.412 sec	9632

This manifold is homeomorphic to the connected sum $L_{3,1} \# L_{3,2}$. See [LD89], [LD91]. Note that since the summands are homeomorphic with distinct orientations, the space is symmetric, hence, the values of the invariant are real. Another blink inducing the same orientable 3-manifold:

TS-code:

$dabcgefihkj$
$khfedcbgjia$
$jiegcbdakhf$

Table 86: Values of the quantum invariants for the 3-manifold induced by $6_3^2(2,0)$

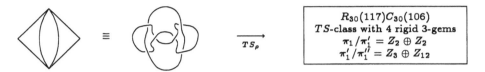

$$R_{30}(117)C_{30}(106)$$
$$TS\text{-class with 4 rigid 3-gems}$$
$$\pi_1/\pi_1' = Z_2 \oplus Z_2$$
$$\pi_1'/\pi_1'' = Z_3 \oplus Z_{12}$$

r	$S^1 \times S^2$ – normalization	S^3 – normalization	C^{++}, 486 time	$r-$ states
3			$0.020sec$	4
4	$+1.00000000 + 0.70710678i$	$+2.00000000 + 1.41421356i$	$0.067sec$	16
5			$0.209sec$	44
6	$+0.28867513 + 1.23205081i$	$+1.00000000 + 4.26794919i$	$0.560sec$	107
7			$1.278sec$	224
8	$-0.66011636 + 1.06066017i$	$-3.44993433 + 5.54327720i$	$2.616sec$	432
9			$4.899sec$	768
10	$-1.21697183 + 0.23434240i$	$-8.80609090 + 1.69571753i$	$8.574sec$	1293

TS-code:
$$dabcgefjhimklon$$
$$jimednhgkabfolc$$
$$nflkhbamodjcegi$$

Table 87: Values of the quantum invariants for the 3-manifold induced by $6_3^2(0,2)$
(another simple projection of Whitehead's link)

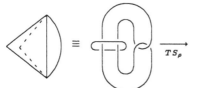

	$R_{26}(11) \in C_{26}(52)$
	TS-class made of
	2 rigid 3-gems
	3-manifold: EUCLID₅

$TS_\rho \uparrow$

r	$S^1 \times S^2 - normalization$	$S^3 - normalization$	$C^{++},486$ time	$r-$ states
3			$0.021sec$	4
4	$-0.38268343 + 0.92387953i$	$-0.76536686 + 1.84775907i$	$0.068sec$	16
5			$0.212sec$	44
6	$-1.00000000i$	$-3.46410162i$	$0.569sec$	107
7			$1.299sec$	224
8	$+0.19509032 + 0.98078528i$	$+1.01959116 + 5.12583090i$	$2.657sec$	432
9			$4.969sec$	768
10	$-0.30901699 - 0.95105652i$	$-2.23606798 - 6.88190960i$	$8.693sec$	1293
11			$14.378sec$	2068
12	$+0.38268343 + 0.92387953i$	$+3.62175489 + 8.74368978i$	$22.742sec$	3184
13			$34.604sec$	4732
14	$-0.43388374 - 0.90096887i$	$-5.15883360 - 10.71242836i$	$51.011sec$	6839
15			$73.162sec$	9632
16	$+0.47139674 + 0.88192126i$	$+6.83432834 + 12.78612901i$	$102.426sec$	13280
17			$140.409sec$	17952
18	$-0.50000000 - 0.86602540i$	$-8.63815572 - 14.96172460i$	$189.096sec$	23865

This manifold can be obtained from a solid cube by identifying via translations two pairs
of opposite faces; the third pair is identified by a $\pi/2$-rotation in one member of the pair
followed by a translation. It is the euclidean 3-manifold whose fundamental group (see
EUCLID₅ in [LS92]) is

$$\langle a,b,c \mid a^4 = b^4 = c^2 = abc \rangle.$$

When $r = 2k + 1$ the quantum invariants seems to vanish. Another blink inducing the
same manifold:

	$cabfdeighkjml$
TS-code:	$imdcbkfjalghe$
	$lgkajhbfeimdc$

Table 88: Values of the quantum invariants the 3-manifold induced by $7_1^2(0,1)$

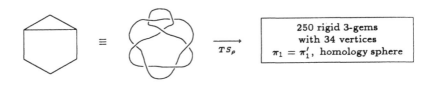

r	$S^1 \times S^2 - normalization$	$S^3 - normalization$	$C^{++}, 486$ time	$r-$ states
3	+0.70710678	+1.00000000	0.020sec	4
4	+0.50000000	+1.00000000	0.044sec	12
5	−0.04387890 + 0.57206140i	−0.11803399 + 1.53884177i	0.105sec	28
6	+0.28867513	+1.00000000	0.219sec	57
7	+0.51544842 + 0.22610588i	+2.22252093 + 0.97492791i	0.410sec	104
8	+0.57402515	+3.00000000	0.706sec	176
9	−0.19777768 + 0.52483389i	−1.22668160 + 3.25519073i	1.141sec	280
10	−0.32532889 + 0.05020285i	−2.35410197 + 0.36327126i	1.756sec	425
11	−0.17361161 + 0.35370761i	−1.44518373 + 2.94434513i	2.591sec	620
12	+0.42264973 + 0.18301270i	+4.00000000 + 1.73205081i	3.696sec	876
13	−0.08469368 + 0.54601809i	−0.90227009 + 5.81691324i	5.131sec	1204
14	−0.34629808 − 0.15676499i	−4.11744901 − 1.86391983i	6.939sec	1617
15	−0.70159629 − 0.17629311i	−9.24142629 − 2.32213284i	9.197sec	2128
16	−0.06517759 − 0.22632747i	−0.94494729 − 3.28130457i	11.963sec	2752
17	−0.26332845 + 0.37366224i	−4.17812119 + 5.92874081i	15.322sec	3504
18	−0.03451354 − 0.10025582i	−0.59626666 − 1.73205081i	19.338sec	4401
19	−0.32392608 − 0.51257414i	−6.06585686 − 9.59849037i	24.101sec	5460
20	+0.31518075 − 0.80125652i	+6.37128803 − 16.19716980i	29.693sec	6700
21	−0.06818995 − 0.15610051i	−1.48253701 − 3.39382573i	36.227sec	8140
22	+0.12832880 + 0.07274939i	+2.99068249 + 1.69541312i	43.758sec	9801
23	−0.13894617 − 0.06668791i	−3.46038765 − 1.66083029i	52.433sec	11704
24	+0.77176762 − 0.34130829i	+20.48233707 − 9.05815583i	62.329sec	13872
25	+0.48976598 − 0.12317374i	+13.81584267 − 3.47461647i	73.600sec	16328

TS-code:
$dabcgefjhimklonqp$
$jipodlqgbacfnmekh$
$qgnjpkocfmiedhbla$

Table 89: Values of the quantum invariants the 3-manifold induced by $7_2^2(-1,0)$

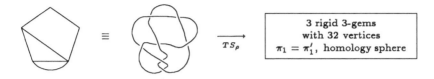

| 3 rigid 3-gems |
| with 32 vertices |
| $\pi_1 = \pi_1'$, homology sphere |

TS_ρ

r	$S^1 \times S^2$ − *normalization*	S^3 − *normalization*	C^{++}, 486 *time*	$r-$ *states*
3	+0.70710678	+1.00000000	0.021*sec*	4
4	+0.50000000	+1.00000000	0.076*sec*	18
5	−0.30075048 + 0.21850801i	−0.80901699 + 0.58778525i	0.237*sec*	52
6	+0.28867513	+1.00000000	0.652*sec*	141
7	+0.58547400 + 0.08069625i	+2.52445867 + 0.34794774i	1.511*sec*	320
8	+1.11522125	+5.82842712	3.190*sec*	680
9	+0.25674180 + 0.26241694i	+1.59239627 + 1.62759536i	6.091*sec*	1312
10	+0.04270510 + 0.13143278i	+0.30901699 + 0.95105652i	10.984*sec*	2405
11	−0.17683775 + 0.24354234i	−1.47203887 + 2.02730357i	18.590*sec*	4148
12	+0.78867513 + 0.18301270i	+7.46410162 + 1.73205081i	30.198*sec*	6882
13	+0.71042338 + 0.44661028i	+7.56837775 + 4.75788856i	47.049*sec*	10948
14	+0.60323536 + 0.32298359i	+7.17240714 + 3.84024209i	71.140*sec*	16905

TS-code:

| *dabcgef jhilknmpo* |
| *jioedclgnaphkbf m* |
| *pjlmckoaf biedgnh* |

Table 90: Values of the quantum invariants of the 3-manifold induced by$7_3^2(3,0)$

r	$S^1 \times S^2$ − *normalization*	S^3 − *normalization*	$C^{++}, 486$ *time*	$r-$ *states*
3	$+1.00000000i$	$+1.41421356i$	$0.026sec$	4
4	$-0.70710678 + 0.70710678i$	$-1.41421356 + 1.41421356i$	$0.147sec$	24
5	$-0.58778525 - 0.80901699i$	$-1.58113883 - 2.17625090i$	$0.628sec$	76
6	$+0.50000000 - 0.86602540i$	$+1.73205081 - 3.00000000i$	$2.165sec$	225
7	$+0.43388374 + 1.34601074i$	$+1.87082869 + 5.80375635i$	$6.009sec$	536
8	$-0.92387953 - 0.38268343i$	$-4.82842712 - 2.00000000i$	$14.544sec$	1184
9	$+0.34202014 - 1.28698898i$	$+2.12132034 - 7.98232487i$	$31.380sec$	2344
10	$+1.30901699 + 0.95105652i$	$+9.47213595 + 6.88190960i$	$62.277sec$	4385
11	$-0.82237337 + 1.71259059i$	$-6.84562884 + 14.25600570i$	$115.115sec$	7676
12	$-1.67303261 - 1.86250130i$	$-15.83375060 - 17.62690155i$	$201.373sec$	12888
13	$+1.53688650 - 0.85460489i$	$+16.37296558 - 9.10439149i$	$335.593sec$	20692
14	$+0.84601074 + 1.75675939i$	$+10.05898175 + 20.88769083i$	$537.866sec$	32193

TS-code: | $dabcgefjhimklonqp$
$jnlpdkhgfoiqbmaec$
$kqojnpicgladfehmb$

Table 91: Values of the quantum invariants the 3-manifold induced by $7^2_4(3,0)$

$$\equiv \qquad \xrightarrow[TS_\rho]{incomplete}$$

| more than 1345 rigid 3-gems |
| with 40 vertices |
| $\pi_1/\pi_1' = Z_3 \oplus Z$ |

r	$S^1 \times S^2 - normalization$	$S^3 - normalization$	$C^{++}, 486$ time	$r-$ states
3	$+1.00000000i$	$+1.41421356i$	$0.020sec$	4
4	$+0.70710678 - 0.70710678i$	$+1.41421356 - 1.41421356i$	$0.049sec$	14
5	$-0.58778525 + 1.42705098i$	$-1.58113883 + 3.83875865i$	$0.121sec$	36
6	$+0.50000000 - 0.86602540i$	$+1.73205081 - 3.00000000i$	$0.274sec$	83
7	$-0.78183148 + 1.17844793i$	$-3.37111682 + 5.08125567i$	$0.538sec$	168
8	$+0.38268343 - 0.92387953i$	$+2.00000000 - 4.82842712i$	$0.989sec$	316
9	$-1.50881301 + 0.26604444i$	$-9.35814981 + 1.65009430i$	$1.685sec$	552
10	$+2.00000000 - 0.72654253i$	$+14.47213595 - 5.25731112i$	$2.740sec$	917
11	$-0.95729268 - 0.34125353i$	$-7.96872883 - 2.84067444i$	$4.248sec$	1452
12	$+0.96592583 + 1.67303261i$	$+9.14162017 + 15.83375060i$	$6.373sec$	2218
13	$-0.40070241 - 1.24851075i$	$-4.26881675 - 13.30080228i$	$9.248sec$	3276
14	$-0.67844793 + 1.40881165i$	$-8.06667705 + 16.75062749i$	$13.101sec$	4711
15	$+0.47706510 - 1.24441485i$	$+6.28390154 - 16.39143240i$	$18.091sec$	6608

Table 92: Values of the quantum invariants the 3-manifold induced by $7_5^2(-3,0)$

		a single 3-gem with 36 vertices $\pi_1/\pi_1' = Z_4$ $\pi_1'/\pi_1'' = Z_5 \oplus Z$

r	$S^1 \times S^2$ – normalization	S^3 – normalization	$C^{++}, 486$ time	$r-$ states
3	$+0.70710678 - 0.70710678i$	$+1.00000000 - 1.00000000i$	$0.013sec$	4
4	$-0.27059805 + 0.65328148i$	$-0.54119610 + 1.30656296i$	$0.049sec$	18
5	$+0.53623297 + 1.41026501i$	$+1.44246348 + 3.79360449i$	$0.159sec$	52
6	$+0.28867513 - 0.50000000i$	$+1.00000000 - 1.73205081i$	$0.438sec$	141
7	$-1.47753438 - 1.09646466i$	$-6.37086268 - 4.72775854i$	$1.019sec$	320
8	$-0.44665909 + 0.12231823i$	$-2.33435292 + 0.63926585i$	$2.144sec$	680
9	$+1.97780684 - 0.63872734i$	$+12.26700233 - 3.96159504i$	$4.112sec$	1312
10	$+1.06524758 - 1.37638192i$	$+7.70820393 - 9.95959314i$	$7.390sec$	2405
11	$-0.56067507 + 1.03888623i$	$-4.66719078 + 8.64793265i$	$12.520sec$	4148
12	$+0.43844775 + 2.08701415i$	$+4.14951408 + 19.75171403i$	$20.305sec$	6882

TS-code:	*dabcgef jhimklpnorq* *jomedlhgqpbrcikanf* *qiogcnrpbkjafmehld*

Table 93: Values of the quantum invariants the 3-manifold induced by $7^2_6(1,0)$

$\equiv$ $\xrightarrow{TS_\rho}$

5 rigid 3-gems with 40 vertices $\pi_1/\pi_1' = Z$

r	$S^1 \times S^2 - normalization$	$S^3 - normalization$	$C^{++}, 486$ time	$r-$ states
3	$+1.00000000$	$+1.41421356$	$0.012sec$	4
4			$0.046sec$	18
5	$+0.50000000 - 0.36327126i$	$+1.34499702 - 0.97719754i$	$0.148sec$	52
6			$0.413sec$	141
7	$-0.59903113 - 1.75675939i$	$-2.58291457 - 7.57483074i$	$0.955sec$	320
8	$+1.00000000 - 1.41421356i$	$+5.22625186 - 7.39103626i$	$2.018sec$	680
9	$+0.28698898 - 0.34202014i$	$+1.77999912 - 2.12132034i$	$3.861sec$	1312
10	$+0.92705098 - 0.95105652i$	$+6.70820393 - 6.88190960i$	$6.965sec$	2405
11	$+0.23259315 + 0.10622173i$	$+1.93615992 + 0.88421461i$	$11.797sec$	4148
12	$-1.00000000 - 2.00000000i$	$-9.46410162 - 18.92820323i$	$19.185sec$	6882
13	$+1.14637228 - 1.84302446i$	$+12.21268710 - 19.63435560i$	$29.924sec$	10948
14	$+0.46950054 - 0.76060704i$	$+5.58231372 - 9.04354053i$	$45.288sec$	16905
15	$+1.54274974 - 1.37786628i$	$+20.32109960 - 18.14925464i$	$66.477sec$	25312
16	$+0.54119610 + 0.22417076i$	$+7.84628224 + 3.25003652i$	$95.427sec$	37024
17	$-0.89049419 - 2.24454461i$	$-14.12909493 - 35.61324087i$	$133.897sec$	52896
18	$+1.48292586 - 1.92836283i$	$+25.61948902 - 33.31499682i$	$184.547sec$	74169
19	$+0.68081427 - 1.11913326i$	$+12.74896398 - 20.95694857i$	$249.778sec$	102084
20	$+2.29360449 - 1.59933924i$	$+46.36455431 - 32.33018218i$	$333.254sec$	138370
21	$+0.85112134 + 0.39887506i$	$+18.50447138 + 8.67205617i$	$438.345sec$	184756
22	$-0.50262429 - 2.42295023i$	$-11.71357953 - 56.46647168i$	$569.838sec$	243573

TS-code:

$dabchefgkijnlmporqts$ $klpedonihqatjrscmgfb$ $mrgiqljbpcfksondthae$

Table 94: Values of the quantum invariants the 3-manifold induced by $7^2_7(3,0)$

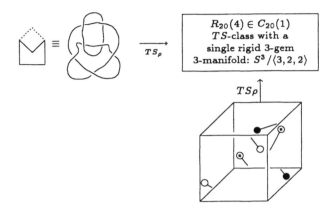

r	$S^1 \times S^2$ − *normalization*	S^3 − *normalization*	$C^{++}, 486$ *time*	$r-$ *states*
3	$+0.70710678 + 0.70710678i$	$+1.00000000 + 1.00000000i$	0.020*sec*	4
4	$-0.65328148 + 0.27059805i$	$-1.30656296 + 0.54119610i$	0.049*sec*	14
5	$-0.01443935 - 1.12968345i$	$-0.03884177 - 3.03884177i$	0.121*sec*	36
6	$+1.15470054$	$+4.00000000$	0.275*sec*	83
7	$-0.40843813 + 0.76906498i$	$-1.76111181 + 3.31606994i$	0.541*sec*	168
8	$-0.55368450 - 0.97130504i$	$-2.89369463 - 5.07628477i$	0.995*sec*	316
9	$+1.07791490 - 0.37080812i$	$+6.68557934 - 2.29987274i$	1.695*sec*	552
10	$-0.17082039 + 0.85065081i$	$-1.23606798 + 6.15536707i$	2.759*sec*	917
11	$-0.70274421 - 0.78095730i$	$-5.84980763 - 6.50087175i$	4.276*sec*	1452
12	$+0.96574120 - 0.48374643i$	$+9.13987282 - 4.57822536i$	6.410*sec*	2218
13	$-0.08661209 + 0.84632810i$	$-0.92270758 + 9.01621614i$	9.312*sec*	3276
14	$-0.74851543 - 0.70449548i$	$-8.89977241 - 8.37637968i$	13.181*sec*	4711
15	$+0.93234396 - 0.52296927i$	$+12.28083457 - 6.88855126i$	18.191*sec*	6608

Smallest blink representing the same orientable 3-manifold:

$T S$-code:
$$\begin{array}{l} cabfdehgji \\ hjdcbifage \\ ifgahbjedc \end{array}$$

Table 95: Values of the quantum invariants the 3-manifold induced by $7^2_8(-3,0)$

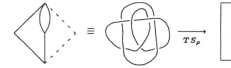

$R_{24}(7) \in C_{24}(37)$
TS-class
with a single 3-gem
EUCLID$_2$, [LS92]

$TS_\rho \uparrow$

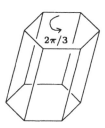

r	$S^1 \times S^2$ − *normalization*	S^3 − *normalization*	C^{++}, 486 time	$r-$ states
3	−1.00000000i	−1.41421356i	0.013sec	4
4			0.049sec	18
5	−0.58778525 − 0.80901699i	−1.58113883 − 2.17625090i	0.158sec	52
6	−0.50000000 + 0.86602540i	−1.73205081 + 3.00000000i	0.437sec	141
7			1.019sec	320
8	+0.38268343 + 0.92387953i	+2.00000000 + 4.82842712i	2.144sec	680
9	+0.64278761 − 0.76604444i	+3.98677815 − 4.75125718i	4.110sec	1312
10			7.390sec	2405
11	−0.28173256 − 0.95949297i	−2.34520788 − 7.98704455i	12.520sec	4148
12	−0.70710678 + 0.70710678i	−6.69213043 + 6.69213043i	20.297sec	6882

TS-code:

cabfdeighljk
ildckgfjaheb
kegjhalbdifc

14.17   Links with 3 Components

Table 96: Values of the quantum invariants for the 3-manifold induced by $6_1^3(0,0,0)$

| | | | | | $R_{24}(28) \in C_{24}(13)$
TS-class with a
single 3-gem
3-manifold: $S^3/\langle 4,3,2 \rangle$ |

r	$S^1 \times S^2 - normalization$	$S^3 - normalization$	$C^{++}, 486$ time	$r-$ states
3			0.041 sec	8
4	$-0.92387953i$	$-1.84775907i$	0.327 sec	48
5			1.960 sec	208
6	$-0.28867513 + 0.86602540i$	$-1.00000000 + 3.00000000i$	8.468 sec	703
7			28.649 sec	2008
8	$+0.31015268 - 0.86572292i$	$+1.62093605 - 4.52448601i$	81.487 sec	5000

Smallest blink inducing the same oriented 3-manifold:

| TS-code: | $dabcgeffjhilk$
$jiledchgkabf$
$kfhjibacelgd$ |

Table 97: Values of the quantum invariants for the 3-torus induced by $6_2^3(0,0,0)$ (the borromean rings)

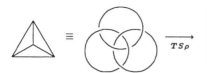

| | | $TS\rho$ | $R_{24}(1) \in C_{24}(40)$
TS-class with a
single 3-gem
3-TORUS |

$TS\rho\uparrow$

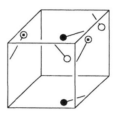

r	$S^1 \times S^2 - normalization$	$S^3 - normalization$	$C^{++}, 486$ time	$r-$ states
3	+2.00000000	+2.82842712	0.019sec	8
4	+3.00000000	+6.00000000	0.106sec	40
5	+4.00000000	+10.75997619	0.462sec	152
6	+5.00000000	+17.32050808	1.563sec	475
7	+6.00000000	+25.87092151	4.395sec	1280
8	+7.00000000	+36.58376302	10.707sec	3072
9	+8.00000000	+49.61860604	23.417sec	6720
10	+9.00000000	+65.12461180	47.095sec	13629
11	+10.00000000	+83.24234537	88.351sec	25960
12	+11.00000000	+104.10511777	156.477sec	46904
13	+12.00000000	+127.84001069	265.411sec	81016
14	+13.00000000	+154.56868004	431.670sec	134615
15	+14.00000000	+184.40799974	678.699sec	216256
16	+15.00000000	+217.47058712	1036.553sec	337280
17	+16.00000000	+253.86523847	1542.887sec	512448
18	+17.00000000	+293.69729464	2240.732sec	760665
19	+18.00000000	+337.06895185	3197.289sec	1105800
20	+19.00000000	+384.07952830	4062.767sec	1577608

TS-code:

$cabfdeighljk$
$ijdclgfkabhe$
$kegjhalbdifc$

Table 98: Values of the quantum invariants for the 3-manifold induced by $6^3_3(0,0,0)$

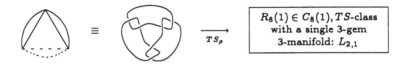

r	$S^1 \times S^2 - normalization$	$S^3 - normalization$	$C^{++}, 486$ time	$r-$ states
3			0.024sec	8
4	-0.38268343	-0.76536686	0.209sec	48
5			1.271sec	208
6	$+0.21132487$	$+0.73205081$	5.508sec	703
7			18.666sec	2008
8	-0.13794969	-0.72095982	53.055sec	5000
9			131.843sec	11256
10	$+0.09893784$	$+0.71592096$	295.621sec	23281
11			609.766sec	45056
12	-0.07535933	-0.71320838	1176.340sec	82352
13			2144.962sec	143624
14	$+0.05984753$	$+0.71158105$	3731.825sec	240399
15			6233.197sec	388624

Smallest blink inducing the same oriented 3-manifold: $\bigwedge$

TS-code:
$$\begin{array}{c} badc \\ dcba \\ cdab \end{array}$$

Table 99: Values of the quantum invariants for the 3-manifold induced by $7^3_1(0,0,-1)$

$R_{28}(172) \in C_{28}(3)$
TS-class made of
6 rigid 3-gems
$\pi_1 = \pi'_1$, homology sphere

r	$S^1 \times S^2 - normalization$	$S^3 - normalization$	$C^{++}, 486$ time	$r-$ states
3	+0.70710678	+1.00000000	0.044*sec*	8
4	−0.50000000	−1.00000000	0.361*sec*	56
5	−0.71637742 − 0.35355339*i*	−1.92705098 − 0.95105652*i*	2.186*sec*	280
6	+0.28867513	+1.00000000	9.654*sec*	1167
7	+0.58547400 + 0.73416118*i*	+2.52445867 + 3.16557105*i*	33.506*sec*	4048
8	+0.19134172*i*	+1.00000000*i*	97.895*sec*	12296

Smallest blink inducing the same oriented 3-manifold:

TS-code:

dabcgefjhilknm
jlmedchgnaibfk
nfikmjaecbdhlg

Bibliography

[Ale30]   J.W. Alexander. The combinatorial theory of complexes. *Annals of Math.*, (2) 31, 1930.

[Art25]   E. Artin. Theorie der zopfe. *Abh. Math. Sem. Univ. Hamburg*, 4:77–72, 1925.

[BL79]   L.C. Biedenharn and J.D. Louck. Angular momentum in quantum physics – theory and application. *(Encyclopedia of Mathematics and Applications)*, 1979.

[CKY93]   L. Crane, L. Kauffman, and D.N. Yetter. Evaluating the Crane Yetter invariant. *(Quantum Topology. edited by Kauffman and Baadhio, World Scientific Press)*, 1993.

[CY93]   L. Crane and D.N. Yetter. A categorical construction of 4d topological quantum field theories. *(Quantum Topology. edited by Kauffman and Baadhio, World Scientific Press)*, 1993.

[Dur92]   C. Durand. Geração e classificação de 3-variedades. Master thesis presented at UFPE, 1992.

[FG82]   Massimo Ferri and Carlo Gagliardi. Crystallization moves. *Pacific J. Math.*, 100:85–103, 1982.

[FG91]   Daniel S. Freed and Robert E. Gompf. Computer calculation of Witten's 3-manifold invariant. *Comm. Math. Phys.*, 141:79–117, 1991.

[FL91]   M. Ferri and S. L. Lins. Topological aspects of edge fusions in 4-graphs. *J. Combin. Th. (B)*, 51:227–243, 1991.

[FR79]   R.A. Fenn and C.P. Rourke. On Kirby's calculus of links. *Topology*, 18:1–15, 1979.

[Jon83]   V.F.R. Jones. Index for subfactors. *Invent. Math.*, 72:1–25, 1983.

[Jon86]   V.F.R. Jones. A polynomial invariant for links via von Neumann algebras. *Bull. Amer. Math. Soc.*, 129:103–112, 1986.

[Jon87]   V.F.R. Jones. Hecke algebra representations of braid groups and link polynomials. *Ann. of Math.*, 126:335–388, 1987.

[Kau83]   L.H. Kauffman. *Formal Knot Theory*, volume 130. Princeton Univ. Press, 1983.

[Kau85]   L.H. Kauffman. State models for link polynomials. *(unpublished notes)*, 1985.

[Kau87a] L.H. Kauffman. On Knots. *Annals Study Number, Princeton University Press*, 115, 1987.

[Kau87b] L.H. Kauffman. State Models and the Jones Polynomial. *Topology*, 26:395–407, 1987.

[Kau89]   L.H. Kauffman. Statistical Mechanics and the Alexander Polynomial. *Contemporary Mathematics*, 96:221–231, 1989.

[Kau90a] L. H. Kauffman. Knots, Spin Networks and Invariants of 3-Manifold. In A. Kawauchi, editor, *KNOTS 90*, pages 271–287, 1990.

[Kau90b] L.H. Kauffman. An invariant of regular isotopy. *Trans. Amer. Math. Soc.*, 318:417–471, 1990.

[Kau90c] L.H. Kauffman. Spin networks and knot polynomials. *Intern. Journal of Modern Physics A.*, 5:93–115, 1990.

[Kau91]   L.H. Kauffman. *Knots and Physics*. World Scientific Pub., 1991.

[Kau92]   L.H. Kauffman. Map coloring, q-deformed spin networks, and Turaev-Viro invariants for 3-manifolds. *Int. J. Mod. Phys. B, Nos. 11,12*, 6:1765–1794, 1992.

[Kir78]   R. Kirby. A calculus for framed links in S^3. *Invent. Math.*, 45:36–56, 1978.

[KM91]   R. Kirby and P. Melvin. On the 3-manifold invariants of Reshetikhin-Turaev for $sl(2, C)$. *Invent. Math.*, 105:473–545, 1991.

[KM92]   B.I. Kurpita and K. Murasugi. On Heirarchical Jones Invariant. In *W. de Gruyter, Pub., (Akio Kawauchi, editor) KNOTS 90*, pages 489–542, 1992.

[KR88]   A.N. Kirillov and N.Y. Reshetikhin. *Representations of the algebra $U_q(sl_2)$, q-orthogonal polynomials and invariants of links. In Infinite Dimensional Lie Algebras and Groups*, volume 7. Ed. by V.G. Kac. Ser. in Math. Phys., 1988.

[LD89]   S. Lins and C. Durand. A complete catalogue of rigid graph-encoded orientable 3-manifolds up to 28 vertices. *Notas Com. Mat. UFPE*, 168, 1989.

[LD91]   S. Lins and C. Durand. Topological classification of small graph-encoded orientable 3-manifolds. *Notas Com. Mat. UFPE*, 177, 1991.

[Lic62]   W.B.R. Lickorish. A representation of orientable combinatorial 3-manifolds. *Ann. of Math.*, 76:531–540, 1962.

[Lic90]   W.B.R. Lickorish. Calculations with the Temperley-Lieb algebra. *Preprint*, 1990.

[Lic91]   W.B.R. Lickorish. 3-manifolds and the Temperley-Lieb algebra. *Math. Annal*, 290:657–670, 1991.

[Lic92]   W.B.R. Lickorish. Skeins and handlebodies. *Preprint*, 1992.

[Lic93]   W.B.R. Lickorish. The Temperley Lieb Algebra and 3-manifold invariants. *Journal of Knot Theory and Its Ramifications*, 2:171–194, 1993.

[Lin82]   S. Lins. Graph-encoded maps. *J. Combin. Th. (B)*, 32:171–181, 1982.

[Lin84]   S. Lins. Towards a catalogue of 3-manifold crystallizations. *Atti. Sem. Mat. Fis. Univ. Modena*, 33:369–378, 1984.

[Lin85]   S. Lins. A simple proof of Gagliardi's handle recognition theorem. *Discrete Math.*, 57:253–260, 1985.

[Lin86]   S. Lins. Paintings: A planar approach to higher dimensions. *GeometriæDedicata*, 20:1–25, 1986.

[Lin88]   S. Lins. On the fundamental group of 3-gems and a "planar" class of 3–manifolds. *Europ. J. Combinatorics*, 9:291–305, 1988.

[LM85]   S. Lins and A. Mandel. Graph-encoded 3-manifolds. *Disc. Math.*, 57:261–284, 1985.

[LS92]   S. Lins and S. Sidki. The fundamental groups of 3-manifolds admitting a colored triangulation with less than 30 tetrahedra. *Submitted*, 1992.

[Mat88]   S. V. Matveev. Transformations of special spines and the Zeeman conjecture. *Math. USSR Izvestia*, 31:2:423–434, 1988.

[Mat91]   S. V. Matveev. Complexity theory of three-dimensional manifolds. *Acta AplicandæMathematicæ, to appear*, 1991.

[Moi77]   E. E. Moise. Graduate texts in mathematics 47. In *Geometric Topology in Dimensions 2 and 3*. Springer-Verlag, 1977.

[Mou79]   J.P. Moussoris. The chromatic evaluation of strand networks. *In Advances in Twistor Theory (edited by Huston and Ward), Research Notes in Mathematics, Pitman Pub.*, pages 308–312, 1979.

[MV92]   G. Masbaum and P. Vogel. 3-valent graphs and the Kauffman bracket. *Preprint*, 1992.

[Nag64]   T. Nagell. *Introduction to Number Theory*. Chelsea Pub. Co., New York, 1964.

[Nei92]   J.R. Neil. Combinatorial calculation of the various normalizations of the Witten invariants for 3-manifolds. *Journal of Knot Theory and Its Ramifications*, 1:407–450, 1992.

[Pen69]   R. Penrose. Angular momentum: an approach to combinatorial space-time. In T.A. Bastin., editor, *Quantum Theory and Beyond*. Cambridge Univ. Press, 1969.

[Pen71]   R. Penrose.   Applications of negative dimensional tensors.   In D. J. A. Welsh, editor, *Combinatorial Mathematics and its Appications*. Academic Press, 1971.

[Pie88]   R. Piergallini.   Standard moves for standard polynedra and spines, III Convegno Nazionale di Topologia Trieste, 9-12 Giugno 1986. In *Supplemento ai Rendiconti del Circolo Matematico di Palermo*, pages 391–414, 1988.

[Piu92]   S. Piunikhin. Turaev-Viro and Kauffman-Lins invariants for 3-manifolds coincide. *Journal of Knot Theory and Its Ramifications*, 1 (2):105–135, 1992.

[Rob93]   J. Roberts. Skein theory and Turaev-Viro invariants. *preprint*, 1993.

[Rol76]   D. Rolfen. *Knots and links*. Publish or Perish Press, 1976.

[Ron93]   Y. Rong. Mutations and Witten Invariants. *preprint*, 1993.

[RT91]   N.Y. Reshetikhin and V. Turaev.   Invariants of three manifolds via link polynomials and quantum groups. *Invent. Math.*, 103:547–597, 1991.

[Ste89]   I. Stewart. *Game, Set and Math*. Blackwell, 1989.

[Tur92]   V.G. Turaev. Topology of Shadows. *Preprint*, 1992.

[TV92]   V.G. Turaev and O.Y. Viro. State sum invariants of 3-manifolds and quantum 6j-symbols. *Topology*, 31:4:865–902, 1992.

[Wit89]   E. Witten.   Quantum field theory and the Jones polynomial. *Commun. Math. Phys.*, 121:351–399, 1989.

Index

3-Vertex, 97
3-gem, 164
TS-moves, 181
θ-Evaluations, 45
θ-Net, 97
n-strand Temperley-Lieb algebra, 8
q-$6j$ Symbols, 99
q-Symmetrizer, 95
(Factorization of 1-dipole cancellation), 169
(Pentagon) Identity, 99
(Recoupling Theorem), 60
(Twist Formula), 24
(bracket state), 11
(vertical) smoothing of σ_i, 11

$(n+1)$-graph, 163
q-deformed factorial, 15
TS_ρ-class, 183
[Orthogonality Identity], 69
[Pentagon Identity (Biedenharn-Elliot)], 71
admissible cycles, 84
backslash crossing, 123
blink, 161
bracket polynomial $< K > \in \mathbb{Z}[A, A^{-1}]$, 6
bundle of loops, 85
elementary tangles, 8
external edges of G, 84

framed isotopy class, 30
gem-complexity, 166
horizontal smoothing of σ_i, 11
internal edges of G, 84
Jones-Wenzl recursion, 91
prime gems, 167
Racah coefficient, 92
ribbon equivalence, 140
Seven Basic Cycles, 89
Seven Cycles, 88
slash crossing, 123
state S of diagram K, 6
switching of a ρ-pair, 176
Temperley-Lieb algebra T_n, 9
trace, 10
writhe, 7

Artin braid group B_n, 11
Axiomatics, 100

Biedenharn-Elliot, 99
Bracket Polynomial, 5
Bubble Move, 104

Chebyschev polynomial, 13
Chromatic Evaluation, 83
Crane-Yetter, 156
Curl and Projector, 96

Dipole-Moves, 167

Edge Dilation, 104

framed isotopic, 30
Framed Links, 129

Gauss sums, 149, 150

Handle Sliding, 133

Jones-Wenzl Projectors, 96

Kirby Calculus, 133

Lickorish's Proof, 144
Loop and Projector, 96
Lune Move, 104

Matveev-Piergallini Moves, 102

Normalization, 146

Orthogonality Identity, 99

Partition Function, 104

Quantum Integers, 95

Recoupling Theorem, 60
Rigid 3-gem, 175

Shadow Interpretation, 153
Shadow Translations, 116
Shadow World, 114
Standard Projector, 13

Tensorial Formalism, 77
Tetrahedral Net, 98
The 3-Vertex, 22
The Reidemeister moves, 7
Turaev-Viro, 102

Vanishing Conditions, 36

Witten-Reshetikhin-Turaev, 129

Y-move, 104

www.ingramcontent.com/pod-product-compliance
Ingram Content Group UK Ltd.
Pitfield, Milton Keynes, MK11 3LW, UK
UKHW020237161224
452563UK00006B/210

9 780691 036403